An Invitation to Mat]

Dierk Schleicher · Malte Lackmann
Editors

An Invitation to Mathematics

From Competitions to Research

Editors
Dierk Schleicher
Jacobs University
Postfach 750 561
D-28725 Bremen
Germany
dierk@jacobs-university.de

Malte Lackmann
Immenkorv 13
24582 Bordesholm
Germany
malte.lackmann@web.de

ISBN 978-3-642-19532-7   e-ISBN 978-3-642-19533-4
DOI 10.1007/978-3-642-19533-4
Springer Heidelberg Dordrecht London New York

Library of Congress Control Number: 2011928905

Mathematics Subject Classification (2010): 00-01, 00A09, 00A05

© Springer-Verlag Berlin Heidelberg 2011
This work is subject to copyright. All rights are reserved, whether the whole or part of the material is concerned, specifically the rights of translation, reprinting, reuse of illustrations, recitation, broadcasting, reproduction on microfilm or in any other way, and storage in data banks. Duplication of this publication or parts thereof is permitted only under the provisions of the German Copyright Law of September 9, 1965, in its current version, and permission for use must always be obtained from Springer. Violations are liable to prosecution under the German Copyright Law.
The use of general descriptive names, registered names, trademarks, etc. in this publication does not imply, even in the absence of a specific statement, that such names are exempt from the relevant protective laws and regulations and therefore free for general use.

Cover design: deblik, Berlin

Printed on acid-free paper

Springer is part of Springer Science+Business Media (www.springer.com)

Contents

Preface: What is Mathematics? vii

Welcome! ix

Structure and Randomness in the Prime Numbers 1
Terence Tao

How to Solve a Diophantine Equation 9
Michael Stoll

From Sex to Quadratic Forms 21
Simon Norton

Small Divisors: Number Theory in Dynamical Systems 43
Jean-Christophe Yoccoz

How do IMO Problems Compare with Research Problems?
Ramsey Theory as a Case Study 55
W. Timothy Gowers

How do Research Problems Compare with IMO Problems?
A Walk Around Games 71
Stanislav Smirnov

Graph Theory Over 45 Years 85
László Lovász

Communication Complexity 97
Alexander A. Razborov

Ten Digit Problems 119
Lloyd N. Trefethen

The Ever-Elusive Blowup in the Mathematical Description of Fluids 137
Robert M. Kerr and Marcel Oliver

About the Hardy Inequality 165
Nader Masmoudi

The Lion and the Christian, and Other Pursuit and Evasion Games 181
Béla Bollobás

Three Mathematics Competitions 195
Günter M. Ziegler

Complex Dynamics, the Mandelbrot Set, and Newton's Method — or: On Useless and Useful Mathematics 207
Dierk Schleicher

Preface: What is Mathematics?

Günter M. Ziegler

This book is an Invitation to Mathematics.

But *What is Mathematics*? This is a question that asks us for a definition. You could look in *Wikipedia* and find the following:

> *Mathematics is the study of quantity, structure, space, and change. Mathematicians seek out patterns, formulate new conjectures, and establish truth by rigorous deduction from appropriately chosen axioms and definitions.*

Quantity, structure, space, and change? These words outline a vast field of knowledge — and they are combined with a very narrow, mechanistic, and, frankly, quite boring description of "what mathematicians do". Should "what mathematicians do" really be a part of the definition?

The definition given by the German *Wikipedia* is interesting in a different way: it stresses that there is no definition of mathematics, or at least no commonly accepted one. I translate:

> *Mathematics is the science that developed from the investigation of figures and computing with numbers. For mathematics, there is no commonly accepted definition; today it is usually described as a science that investigates abstract structures that it created itself for their properties and patterns.*

Is this a good definition, a satisfactory answer to the question "What is Mathematics"? I believe that *Wikipedia* (in any language) does not give a satisfactory answer. At the same time, and much more importantly, high school curricula do not give a satisfactory answer. Even the famous book by Richard Courant and Herbert Robbins entitled "What is Mathematics?" (and subtitled "An Elementary Approach to Ideas and Methods") does not give a satisfactory answer.

Günter M. Ziegler
Fachbereich Mathematik und Informatik, Freie Universität Berlin, Arnimallee 2, 14195 Berlin, Germany. e-mail: ziegler@math.fu-berlin.de

Perhaps it is impossible to give a good definition in a sentence or two. Indeed, I claim that there cannot be one single answer that we could be content with: mathematics in the 21-st century is a huge body of knowledge and a very diverse area of study. There are thus so many ways to experience mathematics — the arenas of national and international competitions, and research experiences that range from years spent working in solitude (think of Andrew Wiles, who proved Fermat's Last Theorem, or Grigori Perelman, who proved the Poincaré conjecture) to coffee break discussions at conferences to massive collaborations on internet platforms (such as the POLYMATH projects initiated by Michael Nielsen, Timothy Gowers, Terence Tao, and others).

But perhaps the English *Wikipedia* is right in one aspect — that in approaching the science called mathematics one should look at the people who do mathematics. So *what is mathematics as an experience*? What does it mean to *do* mathematics?

This book is an invitation to mathematics comprised of contributions by leading mathematicians. Many of them were initiated to mathematics, and led to mathematics research, through competitions such as the mathematical olympiads — one of the ways to get attracted to and drawn into mathematics. This book builds a link between the "domesticated" mathematics taught at high schools or used in competitions and the "wild" and "free" world of mathematical research. As a former high school student, successful participant at competitions such as the IMO 1981, and now professor of mathematics who is doing research and who is active in communicating mathematics to the public, I have personally experienced all these kinds of mathematics, and I am excited about this book and the link that it provides.

The starting point of this book was an event that I had the pleasure of hosting (jointly with Martin Grötschel), namely the 50-th International Mathematical Olympiad, held in Bremen in 2009, at which several premier IMO gold medal winners got on stage to talk about the mathematics that they studied, the mathematics that they are studying, and the mathematics that they are interested in.

All this is reflected in this volume, which contains some of these IMO presentations, as well as other facets of the mathematics research experience. It was put together with admirable care, energy, and attention to detail by Dierk Schleicher (one of the chief organizers of the 50-th IMO in Bremen) and Malte Lackmann (a successful three-time IMO participant). Let me express my gratitude to both of them for this volume, which I see as a book-length exposition of an answer to the question "What is Mathematics?" — and let me wish you an informative, enjoyable, and (in the proper sense of the word) *attractive* reading experience.

Berlin, November 2010

Günter M. Ziegler

Welcome!

Dear Readers,

we are pleased that you have accepted our *Invitation to Mathematics*. This is a joint invitation by a number of leading international mathematicians, together with us, the editors. This book contains fourteen individual invitations, written by different people in different styles, but all of us, authors and editors alike, have one thing in common: we have a passion for mathematics, we enjoy being mathematicians, and we would like to share that enjoyment with you, our readers.

Whom is this book written for? Broadly speaking, this book is written for anyone with an interest in mathematics — yes, for people just like you. More specifically, we have in mind young students at high schools or universities who know mathematics through their classes and possibly through mathematics competitions that they have participated in, either on a local level or all the way up to the level of international olympiads. Mathematics has different flavors: the kind of mathematics found at high school is distinctly different from that found at competitions and olympiads, and both are quite different from mathematics at the research level. Of course, there are similarities too — after all, it's all mathematics we're talking about.

The idea of this book is to allow professional research mathematicians to share their experience and some aspects of their mathematical thinking with our readers. We made a serious effort to reach out to you and write at a level that, for the most part, should be accessible to talented and, most importantly, interested young students in their final years of high school and beyond. Quite importantly, this book is also meant to address high school math teachers, in the hope that they find our invitation interesting as well and share this interest with their students. And of course, we hope that even active research mathematicians will find this book inspiring and, in reading it, will gain new insights into areas outside of their specialization — just as we learned quite a bit of mathematics ourselves in the process of editing this book.

Fourteen invitations to mathematics. You will find that the individual invitations in this book are as varied as the personalities of their authors and their mathematical tastes and preferences: mathematics is a place for very different people. All of our fourteen invitations are independent from each other, and you are welcome to browse through them and start with those that you like best, or that you find most accessible, and continue in your preferred order — much as the white "random path" on the book cover that connects the pictures from the different contributions. (We ordered the contributions by attempting to bring together those that discuss related topics, but of course there would have been many equally good ways to order them.) If you get stuck at one particular invitation, you may want to continue with another. You may discover, though, that some of those that you found difficult at first may become easier, and perhaps more beautiful, once you've learned some more mathematics, or gotten more experience with other invitations, or simply had some time to digest them. Indeed, a number of our contributions are invitations to active reading, and ask you to think along seriously: after all, thinking is what most of us do for much of our professional research time.

Although we encourage you to start by reading those invitations that you prefer, we would like to equally strongly encourage you to fully use the opportunity that this book provides, with the broad area of mathematics that it covers. In high school or during competitions and olympiads, you may have developed preferences for some areas of mathematics or dislikes for others; but mathematics is a broad and rich field, and it is dangerous to become specialized too early, before having seen the beauty of many other mathematical areas. We have often spoken to young university students who thought they were sure they wanted to work in area X and thus refused to take courses in other areas, and advised them to get at least some background in areas Y and Z. Often enough, it turned out that these areas weren't so bad after all, and the students ended up doing their research in areas Y' or Z', or possibly Ω. And even for those few who, after having explored different branches of mathematics, ended up working in exactly the area X that they liked best as a young student, it can only be good to take along as many ideas from other areas of mathematics as possible. In modern mathematics, there is increasingly more interaction between different branches that seemed to be drifting apart some time ago. This can be seen quite well in the articles of our book: many contributions cover (apparently) quite different aspects of mathematical research and show surprising links between them. In addition, there are many links between the different contributions, so that you will often have the feeling of meeting similar ideas in quite different contexts. Rather than telling you where, we encourage you to read and discover this on your own.

To paraphrase the spirit of the book, the title is *not* "Fourteen invitations to mathematics", but "An invitation to mathematics", and we hope that you can get a glimpse of mathematics that has as much breadth and variety as we managed to fit between the covers of this book. (Mathematics itself is much broader, of course, and we thought of many further contributions.

Welcome! xi

If you feel that an important aspect of mathematics is missing here, and that we overlooked a particular person that should have contributed another invitation to mathematics, then please let us know — and help us convince this person to share this invitation with us for the next edition of this book!)

The inspiration for this book. This book was inspired by the 50-th International Mathematical Olympiad that took place in 2009 in Bremen, Germany. Both of us were quite involved in this olympiad: one as a senior organizer, the other as a participant.

One highlight of this olympiad was the 50-th IMO anniversary celebration ceremony, to which six of the world's leading international research mathematicians were invited, all of whom had personal experience with the IMO: Béla Bollobás, Timothy Gowers, László Lovász, Stanislav Smirnov, Terence Tao, and Jean-Christophe Yoccoz. *All six accepted our invitations!* They gave wonderful presentations and were celebrated by the IMO contestants and delegates like movie stars. We tried to provide ample opportunity for IMO contestants and delegates to get in contact with our guests of honor and to have a chance to interact personally with them. This was a most memorable and exciting event that created lasting memories for all of us. We hope that this spirit of personal interaction and invitation will also shine through in this book and its individual contributions.

In addition to the contributions of these guests of honor, three more of our invitations have their roots at the IMO 2009: over the course of three evenings, while the solutions of the contestants were being evaluated, we offered mathematics talks to them (given by Michael Stoll, Marcel Oliver, and Dierk Schleicher). Another contribution (by Alexander Razborov) is based on a lecture series given at the "Summer School on Contemporary Mathematics" held in Dubna/Moscow in 2009. Whatever their inspirations, all contributions were written specifically for this occasion (earlier versions of the contributions by Bollobás, Gowers, Lovász, Smirnov, Tao, and Yoccoz appeared in the report of the 50-th IMO).

This book goes far beyond a single event, exciting as it was, and tries to build lasting links between high schools, competitions, and mathematical research. To use a metaphor of József Pelikán, chairman of the IMO advisory board, research mathematics is like wildlife in uncharted territory, whereas olympiad problems are like animals in a zoo: even though they are presented as animals from the wild, they are constrained to a very restrictive cage. No lion can show its full splendor and strength in the few square meters enclosed by its cage, just as mathematics cannot show its full beauty within the rigid boundaries of competition rules. For young students who have been successful at olympiads, it is important that they learn to leave the olympiad microcosm, to get used to dealing with real mathematical wildlife, and to accept new challenges.

Advertising mathematics, or being a mathematician. We thought about using this introduction to advertise mathematics, including a recitation of the usual claims about how important mathematics is and how much our culture is built upon mathematical thinking. However, we believe that our readers do not need to be convinced, and that the invitations speak for the beauty and value of mathematics by themselves. Nevertheless, we are aware that many students have parents or counselors who tell them that they should study something that will one day earn them money or that has safer job prospects. To them, we would like to say that young people will be most successful in areas that they enjoy the most, because it is only there that they can develop their full potential. Parents[1], please don't worry: all the students from various countries who wanted to become mathematicians and that we advised to pursue their goals despite the concerns of their parents have become quite successful in their fields, in academia, in industry, or in business, and none of them went unemployed.

What makes this book special. First and foremost, our authors include some of the world's leading mathematicians, who are sharing some of their mathematics with you, our readers. This book wants to build a bridge between active research mathematicians and young students; it was realized by a team of people that come from both ends of this bridge: authors, editors, and test readers.

Indeed, we have not made it easy for our authors to write their contributions: we adopted an editing style that Timothy Gowers, in the preface to his *Princeton Companion to Mathematics*, describes as "active interventionist editing". All contributions have been carefully read by us and by a team of young test readers at the age of the intended readership, and we or the authors improved whatever our team could not understand, until things became clear. In this way, we hope that contributions that were *meant to be* comprehensible to our readers actually *are*: the only way to find out was by asking a number of test readers, and that's what we did.

This resulted in numerous and substantial requests for changes in most contributions. All authors accepted these requests, and many were extremely pleased with the feedback they received from us. One author, who was initially somewhat skeptical about this process, wrote "I am extremely impressed by the quality of the job they have done — it greatly surpasses the average level of referee reports I have seen in all three major capacities (editor, author or, well, referee)". In the preface to his *Princeton Companion*, Timothy Gowers writes "given that interventionist editing of this type is rare in mathematics, I do not see how the book can fail to be unusual in a good way". With due modesty, we hope that this applies to some extent to this book as well, and

---

[1] Additional evidence for parents: just a year ago, the *Wall Street Journal* published a ranking of 200 jobs according to five important criteria: work environment, income, employment outlook, physical demands, and stress. The jobs investigated included such different occupations as computer programmer, motion picture editor, physicist, astronomer, and lumber jack. What are the top three jobs? In order, they are: mathematician, actuary, and statistician. All three jobs are based on a strong mathematics education. (Source: http://online.wsj.com/article/SB123119236117055127.html.)

that our readers will appreciate the outcome of the substantial efforts of our authors and our editorial team — our test readers, at least, told us many times that they did.

We would like to conclude this *Welcome* with quotations from two more of our test readers: "I never thought that the topic XY could be exciting to read; well, now I know, it can be!" Another one wrote, after reading a different contribution: "I really found this text very interesting to read; and this really means something because this is not an area I thought I was interested in!"

This is the spirit in which we would like to encourage you to read this book.

Bremen, November 2010

Malte Lackmann and Dierk Schleicher

Acknowledgements. First and foremost, we are indebted to the authors of our *Invitation to Mathematics*. Their willingness to provide their contributions and to share their personal insights, as well as their positive attitude with which they responded to our numerous requests for improvements, are greatly appreciated by us and, we hope, also by our readers. We had a number of "test readers" from the target group of students who patiently and carefully read through some or even all of the invitations, sometimes in several versions, and who helped the authors and us produce a much better book. Several of our authors specifically asked us to convey their appreciation to our test readers for their dedication and care, and we do this with great happiness and gratitude. Our most active test readers were Alexander Thomas, Bertram Arnold, and Kęstutis Česnavičius, but many more students read one or several texts and gave us valuable feedback, including Bastian Laubner, Christoph Kröner, Dima Dudko, Florian Tran, Jens Reinhold, Lisa Sauermann, Matthias Görner, Michael Meyer, Nikita Selinger, Philipp Meerkamp, and Radoslav Zlatev, as well as our colleagues and friends Marcel Oliver and Michael Stoll. We would also like to thank Jan Cannizzo, who greatly helped us take care of English language issues in the texts, and who language edited several contributions entirely; our authors specifically asked us to thank him sincerely. We are extremely grateful to Clemens Heine from Springer Verlag for his untiring and continuous factual and moral support in all kinds of circumstances; if it is ever true to say that a book would not have come into existence without the continuous support of the publishing editor, it is the case here. It has also been a pleasure to work with Frank Holzwarth from Springer Verlag who solved all our LaTeX issues in an instant.

We gratefully acknowledge advice and suggestions shared with us by many colleagues, including Béla Bollobás, Timothy Gowers, Martin Grötschel, and most importantly Günter Ziegler. We would like to thank all those who made the IMO 2009 a success and an inspiration, first and foremost our friends and colleagues in the IMO 2009 steering committee: Anke Allner, Hans-Dietrich Gronau, Hanns-Heinrich Langmann, and Harald Wagner. Moreover, the IMO 2009 had a large team of active helpers: coordinators, team guides, volunteers, and many more — not to forget the many sponsors! We would like to thank them all.

Finally, M.L. would like to thank everyone who made the last year in Bremen possible and, above all, such an enjoyable time.

D.S. would like to thank his students and colleagues for their understanding when he was sometimes preoccupied while editing this book. And of course thank you, Anke and Diego, for your support and understanding all along, and for being with me.

Structure and Randomness in the Prime Numbers

Terence Tao

Abstract. We give a quick tour through some topics in analytic prime number theory, focusing in particular on the strange mixture of order and chaos in the primes. For instance, while primes do obey some obvious patterns (e.g. they are almost all odd), and have a very regular asymptotic distribution (the prime number theorem), we still do not know a deterministic formula to quickly generate large numbers guaranteed to be prime, or to count even very simple patterns in the primes, such as twin primes $p, p+2$. Nevertheless, it is still possible in some cases to understand enough of the structure and randomness of the primes to obtain some quite nontrivial results.

1 Introduction

The prime numbers $2, 3, 5, 7, \ldots$ are one of the oldest topics studied in mathematics. We now have a lot of intuition as to how the primes *should* behave, and a great deal of confidence in our conjectures about the primes... but we still have a great deal of difficulty in *proving* many of these conjectures! Ultimately, this is because the primes are believed to behave *pseudorandomly* in many ways, and not to follow any simple pattern. We have many ways of establishing that a pattern exists... but how does one demonstrate the *absence* of a pattern?

In this article I will try to convince you why the primes are believed to behave pseudorandomly, and how one could try to make this intuition rigorous. This is only a small sample of what is going on in the subject; I am omitting many major topics, such as sieve theory or exponential sums, and am glossing over many important technical details.

Terence Tao
Department of Mathematics, UCLA, Los Angeles CA 90095-1555, USA.
e-mail: tao@math.ucla.edu

2 Finding Primes

It is a paradoxical fact that the primes are simultaneously very numerous, and hard to find. On the one hand, we have the following ancient theorem [2]:

Theorem 1 (Euclid's Theorem). *There are infinitely many primes.*

In particular, given any k, there exists a prime with at least k digits. But there is no known *quick* and *deterministic* way to locate such a prime! (Here, "quick" means "computable in a time which is polynomial in k".) In particular, there is no known (deterministic) formula that can quickly generate large numbers that are guaranteed to be prime. Currently, the largest known prime is $2^{43,112,609} - 1$, about 13 million digits long [3].

On the other hand, one can find primes quickly by *probabilistic* methods. Indeed, any k-digit number can be tested for primality quickly, either by probabilistic methods [10, 12] or by deterministic methods [1]. These methods are based on variants of Fermat's little theorem, which asserts that $a^n \equiv a \bmod n$ whenever n is prime. (Note that $a^n \bmod n$ can be computed quickly, by first repeatedly squaring a to compute $a^{2^j} \bmod n$ for various values of j, and then expanding n in binary and multiplying the indicated residues $a^{2^j} \bmod n$ together.)

Also, we have the following fundamental theorem [8, 14, 16]:

Theorem 2 (Prime Number Theorem). *The number of primes less than a given integer n is $(1 + o(1))\frac{n}{\log n}$, where $o(1)$ tends to zero as $n \to \infty$.*

(We use log to denote the natural logarithm.) In particular, the probability of a randomly selected k-digit number being prime is about $\frac{1}{k \log 10}$. So one can quickly find a k-digit prime with high probability by randomly selecting k-digit numbers and testing each of them for primality.

Is Randomness Really Necessary? To summarize: We do not know a quick way to find primes *deterministically*. However, we have quick ways to find primes *randomly*.

On the other hand, there are major conjectures in complexity theory, such as P = BPP, which assert (roughly speaking) that any problem that can be solved quickly by probabilistic methods can also be solved quickly by deterministic methods.[1]

These conjectures are closely related to the more famous conjecture P ≠ NP, which is a USD $ 1 million Clay Millennium prize problem.[2]

---

[1] Strictly speaking, the P = BPP conjecture only applies to *decision problems* — problems with a yes/no answer —, rather than *search problems* such as the task of finding a prime, but there are variants of P = BPP, such as P = promise-BPP, which would be applicable here.

[2] The precise definitions of P, NP, and BPP are quite technical; suffice to say that P stands for "polynomial time", NP stands for "non-deterministic polynomial time", and BPP stands for "bounded-error probabilistic polynomial time".

Many other important probabilistic algorithms have been *derandomised* into deterministic ones, but this has not been done for the problem of finding primes. (A massively collaborative research project is currently underway to attempt this [11].)

3 Counting Primes

We've seen that it's hard to get a hold of any single large prime. But it is easier to study the set of primes *collectively* rather than one at a time.

An analogy: it is difficult to locate and count all the grains of sand in a box, but one can get an estimate on this count by *weighing* the box, subtracting the weight of the empty box, and dividing by the average weight of a grain of sand. The point is that there is an easily measured statistic (the weight of the box with the sand) which reflects the *collective* behaviour of the sand.

For instance, from the *fundamental theorem of arithmetic* one can establish *Euler's product formula*

$$\sum_{n=1}^{\infty} \frac{1}{n^s} = \prod_{p \text{ prime}} \left(1 + \frac{1}{p^s} + \frac{1}{p^{2s}} + \frac{1}{p^{3s}} + \cdots\right) = \prod_{p \text{ prime}} \left(1 - \frac{1}{p^s}\right)^{-1} \quad (1)$$

for any $s > 1$ (and also for other complex values of s, if one defines one's terms carefully enough).

The formula (1) links the collective behaviour of the primes to the behaviour of the *Riemann zeta function*

$$\zeta(s) := \sum_{n=1}^{\infty} \frac{1}{n^s},$$

thus

$$\prod_{p \text{ prime}} \left(1 - \frac{1}{p^s}\right) = \frac{1}{\zeta(s)}. \quad (2)$$

One can then deduce information about the primes from information about the zeta function (and in particular, its zeroes).

For instance, from the divergence of the harmonic series $\sum_{n=1}^{\infty} \frac{1}{n} = +\infty$ we see that $\frac{1}{\zeta(s)}$ goes to zero as s approaches 1 (from the right, at least). From this and (2) we already recover Euclid's theorem (Theorem 1), and in fact obtain the stronger result of Euler that the sum $\sum_p \frac{1}{p}$ of reciprocals of primes diverges also.[3]

In a similar spirit, one can use the techniques of complex analysis, combined with the (non-trivial) fact that $\zeta(s)$ is never zero for $s \in \mathbb{C}$ when

---

[3] Observe that $\log 1/\zeta(s) = \log \prod_p (1 - p^{-s}) = \sum_p \log(1 - p^{-s}) \geq -2 \sum_p p^{-s}$.

$\mathrm{Re}(s) \geq 1$, to establish the prime number theorem (Theorem 2) [16]; indeed, this is how the theorem was originally proved [8, 14] (and one can conversely use the prime number theorem to deduce the fact about the zeroes of ζ).

The famous *Riemann hypothesis* asserts that $\zeta(s)$ is never zero when[4] $\mathrm{Re}(s) > 1/2$. It implies a much stronger version of the prime number theorem, namely that the number of primes less than an integer $n > 1$ is given by the more precise formula[5] $\int_0^n \frac{dx}{\log x} + O(n^{1/2} \log n)$, where $O(n^{1/2} \log n)$ is a quantity which is bounded in magnitude by $Cn^{1/2} \log n$ for some absolute constant C (for instance, one can take $C = \frac{1}{8\pi}$ once n is at least 2657 [13]). The hypothesis has many other consequences in number theory; it is another of the USD $ 1 million Clay Millennium prize problems. More generally, much of what we know about the primes has come from an extensive study of the properties of the Riemann zeta function and its relatives, although there are also some questions about primes that remain out of reach even assuming strong conjectures such as the Riemann hypothesis.

4 Modeling Primes

A fruitful way to think about the set of primes is as a *pseudorandom set* — a set of numbers which is not actually random, but behaves like one.

For instance, the prime number theorem asserts, roughly speaking, that a randomly chosen large integer n has a probability of about $1/\log n$ of being prime. One can then *model* the set of primes by replacing them with a random set of integers, in which each integer $n > 1$ is selected with an independent probability of $1/\log n$; this is *Cramér's random model*.

This model is too crude, because it misses some obvious structure in the primes, such as the fact that most primes are odd. But one can improve the model to address this, by picking a model where odd integers n are selected with an independent probability of $2/\log n$ and even integers are selected with probability 0.

One can also take into account other obvious structure in the primes, such as the fact that most primes are not divisible by 3, not divisible by 5, etc. This leads to fancier random models which we believe to accurately predict the asymptotic behaviour of primes.

---

[4] A technical point: the sum $\sum_{n=1}^{\infty} \frac{1}{n^s}$ does not converge in the classical sense when $\mathrm{Re}(s) \leq 1$, so one has to interpret this sum in a fancier way, or else use a different definition of $\zeta(s)$ in this case; but I will not discuss these subtleties here.

[5] The Prime Number Theorem in the version of Theorem 2 says that, as $n \to \infty$, the number of correct decimal digits in the estimate $n/\log n$ tends to infinity, but it does not relate the number of correct digits to the total number of digits of $\pi(n)$. If the Riemann hypothesis is correct, then $\int_0^n dx/\log x$ correctly predicts almost half of the digits in $\pi(n)$.

For example, suppose we want to predict the number of twin primes n, $n+2$, where $n \leq N$ for a given threshold N. Using the Cramér random model, we expect, for any given n, that $n, n+2$ will simultaneously be prime with probability $\frac{1}{\log n \log(n+2)}$, so we expect the number of twin primes to be about[6]

$$\sum_{n=1}^{N} \frac{1}{\log n \log(n+2)} \approx \frac{N}{\log^2 N}.$$

This prediction is inaccurate; for instance, the same argument would also predict plenty of pairs of *consecutive* primes $n, n+1$, which is absurd. But if one uses the refined model where odd integers n are prime with an independent probability of $2/\log n$ and even integers are prime with probability 0, one gets the slightly different prediction

$$\sum_{\substack{1 \leq n \leq N \\ n \text{ odd}}} \frac{2}{\log n} \times \frac{2}{\log(n+2)} \approx 2 \frac{N}{\log^2 N}.$$

More generally, if one assumes that all numbers n divisible by some prime less than a small threshold w are prime with probability zero, and are prime with a probability of $\prod_{p<w}(1-\frac{1}{p})^{-1} \times \frac{1}{\log n}$ otherwise, one is eventually led to the prediction

$$2 \left(\prod_{\substack{p<w \\ p \text{ odd}}} \frac{p-2}{p} \left(1-\frac{1}{p}\right)^{-2} \right) \frac{N}{\log^2 N} = 2 \left(\prod_{\substack{p<w \\ p \text{ odd}}} \left(1-\frac{1}{(p-1)^2}\right) \right) \frac{N}{\log^2 N}$$

(for p an odd prime, among p consecutive integers, only $p-2$ have a chance to be the smaller number in a pair of twin primes). Sending $w \to \infty$, one is led to the asymptotic prediction

$$\Pi_2 \frac{N}{\log^2 N}$$

for the number of twin primes less than N, where Π_2 is the *twin prime constant*

$$\Pi_2 := 2 \prod_{p \text{ odd prime}} \left(1 - \frac{1}{(p-1)^2}\right) \approx 1.32032\ldots.$$

For $N = 10^{10}$, this prediction is accurate to four decimal places, and is believed to be asymptotically correct. (This is part of a more general conjecture, known as the *Hardy-Littlewood prime tuples conjecture* [9].)

---

[6] We use the symbol $\approx$ in the sense that the quotient of the two quantities tends to 1 as $N \to \infty$.

Similar arguments based on random models give convincing heuristic support for many other conjectures in number theory, and are backed up by extensive numerical calculations.

5 Finding Patterns in Primes

Of course, the primes are a deterministic set of integers, not a random one, so the predictions given by random models are not rigorous. But can they be made so?

There has been some progress in doing this. One approach is to try to classify all the possible ways in which a set could *fail* to be pseudorandom (i.e. it does something noticeably different from what a random set would do), and then show that the primes do not behave in any of these ways.

For instance, consider the *odd Goldbach conjecture*: every odd integer larger than five is the sum of three primes. If, for instance, all large primes happened to have their last digit equal to one, then Goldbach's conjecture could well fail for some large odd integers whose last digit was different from three. Thus we see that the conjecture could fail if there was a sufficiently strange "conspiracy" among the primes.

However, one can rule out this particular conspiracy by using the *prime number theorem in arithmetic progressions*, which tells us that (among other things) there are many primes whose last digit is different from 1. (The proof of this theorem is based on the proof of the classical prime number theorem.)

Moreover, by using the techniques of *Fourier analysis* (or more precisely, the *Hardy-Littlewood circle method*), we can show that *all* the conspiracies which could conceivably sink Goldbach's conjecture (for large integers, at least) are broadly of this type: an unexpected "bias" for the primes to prefer one remainder modulo 10 (or modulo another base, which need not be an integer), over another.

Vinogradov [15] eliminated each of these potential conspiracies, and established *Vinogradov's theorem*: every sufficiently large odd integer is the sum of three primes.[7] This method has since been extended by many authors, to cover many other types of patterns; for instance, related techniques were used by Ben Green and myself [4] to establish that the primes contain arbitrarily long arithmetic progressions, and in subsequent work of Ben Green, myself, and Tamar Ziegler [5, 6, 7] to count a wide range of other additive patterns also. (Very roughly speaking, known techniques can count additive patterns that involve two independent parameters, such as arithmetic progressions $a, a+r, \ldots, a+(k-1)r$ of a fixed length k.)

---

[7] Vinogradov himself could not specify explicitly what "sufficiently large" is. Soon after, his student Borozdin showed that numbers greater than $3^{3^{15}} \approx 10^{6\,846\,169}$ are "sufficiently large". Meanwhile, this bound has been lowered to $e^{3\,100} \approx 10^{1\,346}$ — still far beyond reach for computer tests for the smaller numbers.

Unfortunately, "one-parameter" patterns, such as twins $n, n+2$, remain stubbornly beyond current technology. There is still much to be done in the subject!

References

[1] Manindra Agrawal, Neeraj Kayal, and Nitin Saxena, PRIMES is in P. *Annals of Mathematics (2)* **160**, 781–793 (2004)
[2] Euclid, *The Elements*, circa 300 BCE
[3] Great Internet Mersenne Prime Search. http://www.mersenne.org (2008)
[4] Ben Green and Terence Tao, The primes contain arbitrarily long arithmetic progressions. *Annals of Mathematics* **167**(2), 481–547 (2008)
[5] Ben Green and Terence Tao, *Linear equations in primes*. Preprint. http://arxiv.org/abs/math/0606088, 84 pages (April 22, 2008)
[6] Ben Green and Terence Tao, *The Möbius function is asymptotically orthogonal to nilsequences*. Preprint. http://arxiv.org/abs/0807.1736, 22 pages (April 26, 2010)
[7] Ben Green, Terence Tao and Tamar Ziegler, *The inverse conjecture for the Gowers norm*. Preprint
[8] Jacques Hadamard, Sur la distribution des zéros de la fonction $\zeta(s)$ et ses conséquences arithmétiques. *Bulletin de la Société Mathématique de France* **24**, 199–220 (1896)
[9] Godfrey H. Hardy and John E. Littlewood, Some problems of 'partitio numerorum'. III. On the expression of a number as a sum of primes. *Acta Mathematica* **44**, 1–70 (1923)
[10] Gary L. Miller, Riemann's hypothesis and tests for primality. *Journal of Computer and System Sciences* **13**(3), 300–317 (1976)
[11] Polymath4 project: Deterministic way to find primes. http://michaelnielsen.org/polymath1/index.php?title=Finding_primes
[12] Michael O. Rabin, Probabilistic algorithm for testing primality. *Journal of Number Theory* **12**, 128–138 (1980)
[13] Lowell Schoenfeld, Sharper bounds for the Chebyshev functions $\theta(x)$ and $\psi(x)$. II. *Mathematics of Computation* **30**, 337–360 (1976)
[14] Charles-Jean de la Vallée Poussin, Recherches analytiques de la théorie des nombres premiers. *Annales de la Société scientifique de Bruxelles* **20**, 183–256 (1896)
[15] Ivan M. Vinogradov, The method of trigonometrical sums in the theory of numbers (Russian). *Travaux de l'Institut Mathématique Stekloff* **10** (1937)
[16] Don Zagier, Newman's short proof of the prime number theorem. *American Mathematical Monthly* **104**(8), 705–708 (1997)

How to Solve a Diophantine Equation

Michael Stoll

Abstract. We introduce Diophantine equations and show evidence that it can be hard to solve them. Then we demonstrate how one can solve a specific equation related to numbers occurring several times in Pascal's Triangle with state-of-the-art methods.

1 Diophantine Equations

The topic of this text is *Diophantine Equations*. A Diophantine equation is an equation of the form

$$F(x_1, x_2, \ldots, x_n) = 0,$$

where F is a polynomial with integer coefficients, and one asks for solutions in *integers* (or rational numbers, depending on the problem). They are named after Diophantos of Alexandria on whom not much is known with any certainty. Most likely he lived around 300 AD. He wrote the *Arithmetika*, a text consisting of 13 books, a number of which have been preserved. In this text, he explains through many examples ways of solving certain kinds of equations like the above in rational numbers. Diophantos was also one of the first to introduce symbolic notation for the powers of an indeterminate.

To give you a flavor of this kind of question, let me show you some examples. Ideally, you should cover up the part of the page below the equation and try to find a solution for yourself before you read on. The first equation

---

Michael Stoll
Mathematisches Institut, Universität Bayreuth, 95440 Bayreuth, Germany.
e-mail: Michael.Stoll@uni-bayreuth.de

is
$$x^3 + y^3 + z^3 = 29,$$
an equation in three unknowns, to be solved in (not necessarily positive) integers. I trust it did not take you very long to come up with a solution like $(x, y, z) = (3, 1, 1)$ or maybe $(4, -3, -2)$. Now let us look at
$$x^3 + y^3 + z^3 = 30.$$
Try to solve it for a while before you look up a solution in this footnote[1]. This solution is the smallest and was found by computer search in July 1999 and published in 2007 [1]. This already indicates that it may be quite hard to find a solution to a given Diophantine equation. Now consider
$$x^3 + y^3 + z^3 = 31.$$
Did you try to solve it? You should have come to the conclusion that there is no solution: the third power of an integer is always $\equiv -1, 0$ or $1 \bmod 9$, so a sum of three cubes can never be $\equiv 4$ or $5 \bmod 9$. Since $31 \equiv 4 \bmod 9$, the number 31 cannot be a sum of three cubes. If we replace 31 with 32, the same argument applies. So we consider
$$x^3 + y^3 + z^3 = 33$$
next. If you were able to solve this, you should consider making Diophantine equations your research area. The sad state of affairs is that it is an open problem whether this equation has a solution in integers or not![2]

So the following looks like an interesting problem: to decide if a given Diophantine equation is solvable or not. In fact, this problem appears on the most famous list of mathematical problems, namely the 23 problems David Hilbert stated in his address to the International Congress of Mathematicians in Paris in 1900 as questions worth working on in the new century. The description of the tenth problem in Hilbert's list reads thus (in the German original [3], see [4] for an English translation of Hilbert's address):

---

10. Entscheidung der Lösbarkeit einer Diophantischen Gleichung.

Eine D i o p h a n t i s c h e Gleichung mit irgend welchen Unbekannten und mit ganzen rationalen Zahlencoefficienten sei vorgelegt: *man soll ein Verfahren angeben, nach welchem sich mittelst einer endlichen Anzahl von Operationen entscheiden läßt, ob die Gleichung in ganzen rationalen Zahlen lösbar ist.*

---

[1] $x = 2\,220\,422\,932, \quad y = -2\,218\,888\,517, \quad z = -283\,059\,965$.
[2] This introduction was inspired by a talk Bjorn Poonen gave at a workshop in Warwick in 2008.

Here is an English translation.

> Given a Diophantine equation with any number of unknown quantities and with rational integral numerical coefficients: *to devise a process according to which it can be determined by a finite number of operations whether the equation is solvable in rational integers.*

In modern terminology, Hilbert asks for an *algorithm* that, given an arbitrary polynomial $F(x_1, \ldots, x_n)$ with integral coefficients, decides whether the equation
$$F(x_1, \ldots, x_n) = 0$$
can be solved in integers. This is commonly known as *Hilbert's Tenth Problem*. It is not only the shortest problem on Hilbert's list, it is also the only decision problem[3], so it is somewhat special. From the wording it can be inferred that Hilbert believed in a positive solution to his problem: such an algorithm had to exist. In fact, at the end of the introductory part of his speech, before turning to the list of problems, he says

> ... in der Mathematik giebt es kein Ignorabimus!

(There is no 'Ignorabimus'[4] in mathematics.) This indicates that Hilbert was convinced that every mathematical problem must have a definite solution.

The simple examples I have shown at the beginning may (or should) have given you a feeling that this problem may actually be very hard. This is also what happened historically. People got more and more convinced that the answer to Hilbert's Tenth Problem was likely to be negative: an algorithm conforming to the given specification does not exist. Now if an algorithm does exist that performs a certain task, it is fairly clear how one can prove this fact. Namely, one has to find such an algorithm and write it down, then everybody will agree that it indeed *is* an algorithm solving the given problem. To show that such an algorithm *does not* exist is a quite different matter. One needs some way of getting a handle on all possible algorithms, so that one can show that none of them solves the problem. The relevant theory, which is a branch of mathematical logic, did not yet exist when Hilbert gave his talk. It was developed a few decades later, leading to such famous results as Gödel's Incompleteness Theorem, which definitely showed that there is an *Ignorabimus* in mathematics. Indeed, work of several people, most notably Martin Davis, Hilary Putnam and Julia Robinson, made it possible for Yuri Matiyasevich to finally prove in 1970 the following result.[5]

Theorem 1 (Davis, Putnam, Robinson; Matiyasevich).
The solvability of Diophantine equations is undecidable.

---

[3] A *decision problem* asks for an algorithm that decides if a given element of a specified set has a specified property.

[4] This Latin word means 'we will not know'.

[5] See [6] for an accessible account of the problem and its solution.

In fact, he proved a much stronger result, which implies for example that there is an explicit polynomial $F(x_0, x_1, \ldots, x_n)$ such that there is no algorithm that, given $a \in \mathbb{Z}$ as input, decides whether or not there is an integral solution to
$$F(a, x_1, \ldots, x_n) = 0.$$
Note that if a Diophantine equation is solvable, then we can prove it, since we will eventually find a solution by searching through the countably many possibilities (but we do not know beforehand how far we have to search). So the really hard problem is to prove that there are no solutions when this is the case. A similar problem arises when there are finitely many solutions and we want to find them all. In this situation one expects the solutions to be fairly small.[6] So usually it is not so hard to find all solutions; what is difficult is to show that there are no others.

So, given Theorem 1, should we give up all attempts to solve Diophantine equations, convinced that the task is completely hopeless? That would be premature. We might still be able to prove positive results when we restrict the set of equations in some way. For example, there are quite good reasons to believe that there should be a positive answer to Hilbert's question for equations *in two variables*. In the remainder of this contribution, we will consider one such equation as an example case and show with what kind of methods it can be attacked and solved.

2 The Example Equation

The equation we want to consider here is motivated by the following question. Consider Pascal's Triangle (Fig. 1). Which natural numbers occur several times in this triangle, if we disregard the outer two "layers" $(1, 1, 1, \ldots$ and $1, 2, 3, \ldots)$ on either side and the obvious reflectional symmetry?

In other words, what are the integral solutions to the equation
$$\binom{y}{k} = \binom{x}{l}, \tag{1}$$
subject to the conditions $1 < k \leq y/2$, $1 < l \leq x/2$ and $k < l$?

---

[6] The large solution to $x^3 + y^3 + z^3 = 30$ is no counterexample to this statement, since there should be infinitely many solutions in this case.

How to Solve a Diophantine Equation

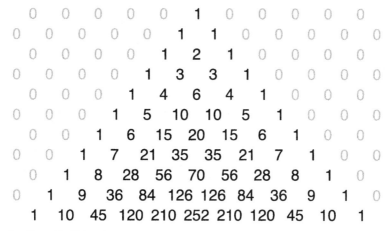

Fig. 1. Pascal's Triangle.

The following solutions are known.

$$\binom{16}{2} = \binom{10}{3}, \quad \binom{56}{2} = \binom{22}{3}, \quad \binom{120}{2} = \binom{36}{3},$$

$$\binom{21}{2} = \binom{10}{4}, \quad \binom{153}{2} = \binom{19}{5}, \quad \binom{78}{2} = \binom{15}{5} = \binom{14}{6},$$

$$\binom{221}{2} = \binom{17}{8}, \quad \binom{F_{2i+2}F_{2i+3}}{F_{2i}F_{2i+3}} = \binom{F_{2i+2}F_{2i+3} - 1}{F_{2i}F_{2i+3} + 1} \text{ for } i = 1, 2, \ldots,$$

where F_n is the n-th Fibonacci number.

Equation (1) is not a Diophantine equation according to our definition, since it depends on k and l in a non-polynomial way. Also, it is way too hard to solve. So we specialize by fixing k and l. The cases

$$(k, l) \in \{(2, 3), (2, 4), (2, 6), (2, 8), (3, 4), (3, 6), (4, 6), (4, 8)\}$$

have already been solved completely [7]. Each of these cases requires some deep mathematics of a flavor similar to what is described below. The next interesting case is obviously $(k, l) = (2, 5)$, leading to the equation

$$\binom{y}{2} = \binom{x}{5}, \quad \text{or} \quad 60y(y - 1) = x(x - 1)(x - 2)(x - 3)(x - 4). \quad (2)$$

The first step in solving an equation like (2) is to go and look for its solutions. We easily find solutions with

$$x = 0, 1, 2, 3, 4, 5, 6, 7, 15, \quad \text{and} \quad 19,$$

and then no further ones. (Only the last two are 'nontrivial' in the sense that they satisfy the constraints given above. Also, there are no solutions with $x < 0$, since then the right hand side is negative, but the left hand side can never be negative for $y \in \mathbb{Z}$.) This now raises the question if we have already found them all, and if so, how to prove it.

This is a good point to look at what is known about the solution set of equations like (2) in general. The first important result was proved by Carl Ludwig Siegel in 1929. (See [5, Section D.9] for a proof.)

Theorem 2 (Siegel).
Let F be a polynomial with integer coefficients in two variables x and y. If the solutions to $F(x, y) = 0$ cannot be rationally parameterized, then $F(x, y) = 0$ has only finitely many solutions in integers.

A *rational parameterization* of $F(x, y) = 0$ is a pair of rational functions $f(t)$, $g(t)$ (quotients of polynomials), not both constant, such that $F(f(t), g(t)) = 0$ (as a function of t). The existence of such a rational parameterization can be algorithmically checked; for our equation it turns out that it is not rationally parameterizable. So we already know that there are only finitely many solutions. In particular, we have a chance that our list is complete. On the other hand, Theorem 2 and its proof are inherently *ineffective*: we do not get a bound on the size of the solutions, so this result gives us no way of checking that our list is complete. This somewhat unsatisfactory state of affairs did not change until the 1960s, when Alan Baker developed his theory of 'linear forms in logarithms' that for the first time provided explicit bounds for solutions of many types of equations. For this breakthrough, he received the Fields Medal. Baker's results cover a class of equations that contains our equation (2). For our case, what he proved comes down to roughly the following:

$$|x| < 10^{10^{10^{10^{600}}}}. \tag{3}$$

This reduces the solution of our equation (2) to a finite problem. The inequality in (3) gives us an explicit upper bound for x. So we only have to check the finitely many possibilities that remain, and we will obtain the complete set of solutions to (2). From a very pure mathematics viewpoint, we may therefore consider our problem as solved. On the other hand, from a more practical point of view, we would like to actually obtain the complete list of solutions, and the assertion that it is possible in principle to get it does not satisfy us. To say that the number showing up in (3) defies all imagination is a horrible understatement, and one cannot even begin to figure out how long it would take to actually perform all the necessary computations.

However, time did not stop in the 1960s, and with basically still the same method, but with a lot of refinements and improvements thrown in, we are now able to prove the following estimate:

$$|x| < 10^{10^{10^{600}}}. \tag{4}$$

You may rightly ask whether something has really been gained, in practical terms. The number of electrons in the universe is estimated to be about 10^{80}, so we cannot even write down a number with something like 10^{600} digits! However, it will turn out that the improvement represented by (4) is crucial. But before we can see this, we need to look at our problem from a different angle.

3 A Geometric Interpretation

The idea is to translate our at first sight algebraic problem ((2) is an algebraic equation) into a geometric one. An equation $F(x, y) = 0$ in two variables defines a subset of the plane, consisting of those points whose coordinates satisfy the equation. If F is a (non-constant) polynomial, this solution set is called a *plane algebraic curve*. We can draw the curve C corresponding to our equation (2) in the real plane $\mathbb{R}^2$, see Fig. 2. We are now interested in the *integral points* on C, since they correspond to integral solutions to (2). The set of integral points on C is denoted $C(\mathbb{Z})$.

This set $C(\mathbb{Z})$ of integral points on the curve C by itself does not have any useful additional structure. But we can make use of a well-developed theory, called Algebraic Geometry, that studies sets defined by a collection of polynomial equations, and in particular algebraic curves like C. This theory tells us that we can *embed* the curve C into another object J, which is not a curve, but a surface. This can be constructed for any curve and is called the *Jacobian variety* of the curve.[7] The interesting fact about J (and Jacobian varieties in general) is that J is a *group*. More precisely, there is a composition law on J that is defined in a geometric way and that turns (for example) the set $J(\mathbb{Z})$ of integral points[8] on J into an *abelian group*. In a similar spirit as Siegel's Theorem 2 (and actually preceding it by one year), we have the following important result by André Weil, valid for Jacobian varieties in general. (See [5, Part C].)

Theorem 3 (Weil).
If J is the Jacobian variety of a curve, then the abelian group $J(\mathbb{Z})$ is finitely generated.

This means that we can (in principle) get an explicit description of the group $J(\mathbb{Z})$ in terms of generators and relations. If we have that, we may be able to use the group structure and the geometry in some way to get a handle on the elements of $J(\mathbb{Z})$ that are in the image of C; these correspond exactly to the integral points on C.

---

[7] The Jacobian variety need not be a surface; its dimension depends on the curve.

[8] Algebraic geometers use the set of *rational* points here. This does not make a difference, since J is a projective variety (this means that the coordinates can be scaled so as to remove denominators).

Fig. 2. The curve given by (2), with some integral points.

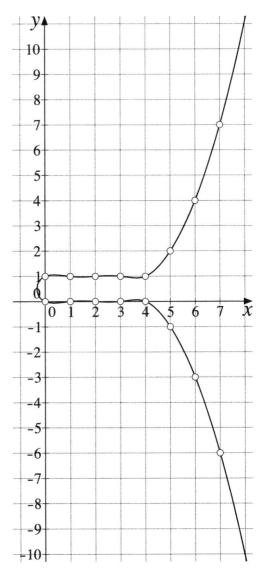

In general, it is not known whether it is always possible to actually determine explicit generators of a group like $J(\mathbb{Z})$ in an algorithmic way, although there are some 'standard conjectures' whose truth would imply a positive answer. There are methods available that, with some luck, can find a set of generators, but they are not guaranteed to work in all cases. This is the point where the method we are describing may fail in practice. In our specific example, we are lucky, and we can show that $J(\mathbb{Z})$ is a *free abelian group of rank 6*:

$$J(\mathbb{Z}) = \mathbb{Z}\, P_1 + \mathbb{Z}\, P_2 + \mathbb{Z}\, P_3 + \mathbb{Z}\, P_4 + \mathbb{Z}\, P_5 + \mathbb{Z}\, P_6 \qquad (5)$$

with explicitly known points $P_1, \ldots, P_6 \in J(\mathbb{Z})$.

Let $\iota : C \to J$ denote the embedding of C into J. The surface J lives in some high-dimensional space, and we can specify integral points on it by a bunch of coordinates. We can measure the size of such a point by taking the logarithm of the largest absolute value of the coordinates (this tells us roughly how much space we need to write the point down). This gives us a function

$$h : J(\mathbb{Z}) \to \mathbb{R}_{\geq 0}$$

called the *height*. One can show that this height function has the following properties. The first one tells us how the height relates to the size of integral points on our curve.

$$h\bigl(\iota(x,y)\bigr) \approx \log |x| \quad \text{for points } (x,y) \in C(\mathbb{Z}) \text{ such that } x \text{ is not very small.} \qquad (6)$$

The second property says that the height function behaves well with respect to the group structure on J.

$$h(n_1 P_1 + n_2 P_2 + n_3 P_3 + n_4 P_4 + n_5 P_5 + n_6 P_6) \approx n_1^2 + n_2^2 + n_3^2 + n_4^2 + n_5^2 + n_6^2. \qquad (7)$$

(To be precise, each side can be bounded by an explicit constant multiple of the other one. To be more precise, h is, up to a bounded error, a positive definite quadratic form on $J(\mathbb{Z})$.)

If we now combine the estimate (4) with the properties (6) and (7) of the height h, then we obtain the following statement.

Lemma 1. *If $(x, y) \in C(\mathbb{Z})$, then we have*

$$\iota(x, y) = n_1 P_1 + n_2 P_2 + n_3 P_3 + n_4 P_4 + n_5 P_5 + n_6 P_6$$

with coefficients $n_j \in \mathbb{Z}$ satisfying $|n_j| < 10^{300}$.

Of course, the bound 10^{300} given here is not precise; a precise bound can be given and is of the same order of magnitude.

The conclusion is that using the additional structure we have on J enables us to reduce the size of the search space from about $10^{10^{600}}$ to 'only' 10^{1800} (there are six coefficients n_j with about 10^{300} possible values each). This is, of course, still much too large to check each possibility (think of the electrons in the universe), but, and this is the decisive point, the numbers n_j we have to deal with can be represented easily on a computer, and we *can compute* with them!

4 Needles in a Haystack

We now have an enormous haystack

$$H = \{(n_1, n_2, n_3, n_4, n_5, n_6) \in \mathbb{Z}^6 : |n_j| < 10^{300}\}$$

of about 10^{1800} pieces of grass that contains a small number of needles. We want to find the needles. Instead of looking at each blade of grass in the haystack, we can try to solve this problem faster by finding conditions on the possible positions of the needles that rule out large parts of the haystack. This is the point where we make use of the fact that the group law on J is defined via geometry. Our objects C, J and ι are defined over $\mathbb{Z}$, therefore it makes sense to consider their defining equations modulo p for prime numbers p. We denote the field $\mathbb{Z}/p\mathbb{Z}$ of p elements by $\mathbb{F}_p$. The sets of points with coordinates in $\mathbb{F}_p$ that satisfy these defining equations mod p are denoted by $C(\mathbb{F}_p)$ and $J(\mathbb{F}_p)$. Then for all but finitely many p (and the exceptions can be found explicitly), $J(\mathbb{F}_p)$ is again an abelian group, and it contains the image $\iota(C(\mathbb{F}_p))$ of $C(\mathbb{F}_p)$. The group $J(\mathbb{F}_p)$ is finite, and so is the set $C(\mathbb{F}_p)$; both can be computed. Furthermore, the following diagram commutes, and the geometric nature of the group structure implies that the right hand vertical map is a group homomorphism.

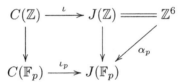

The vertical maps are obtained by reducing the coordinates of the points mod p. The diagonal map α_p is again a group homomorphism, determined by the image of the generators $P_1, \ldots, P_6$ of $J(\mathbb{Z})$. The following is now clear.

Lemma 2. *Let $(x, y) \in C(\mathbb{Z})$ and $\iota(x, y) = n_1 P_1 + \cdots + n_6 P_6$. Then*

$$\alpha_p(n_1, n_2, n_3, n_4, n_5, n_6) \in \iota_p(C(\mathbb{F}_p)).$$

The subset $\Lambda_p = \alpha_p^{-1}(\iota_p(C(\mathbb{F}_p))) \subset \mathbb{Z}^6$ is (usually, when α_p is surjective) a union of $\#C(\mathbb{F}_p)$ cosets of a subgroup of index $\#J(\mathbb{F}_p)$ in $\mathbb{Z}^6$. Since one can show that $\#C(\mathbb{F}_p) \approx p$ and $\#J(\mathbb{F}_p) \approx p^2$ (reflecting dimensions 1 and 2, respectively), we see that the intersection of our haystack H with Λ_p has only about $1/p$ times as many elements as the original haystack. This does not yet help very much, but we can try to *combine* information coming from *many* primes. If S is a (finite, but large) set of prime numbers, then we set

$$\Lambda_S = \bigcap_{p \in S} \Lambda_p \quad \text{and obtain} \quad \iota(C(\mathbb{Z})) \subset \Lambda_S \cap H.$$

If we make S sufficiently large (about a thousand primes, say), then it is likely that the set on the right hand side is quite small, so that we can easily check the remaining possibilities. The idea is that the reductions of the haystack size we obtain from several distinct primes should accumulate, so that we can expect a reduction by a factor which is roughly the product of all the primes in S.

We have to be careful to select the primes in a good way so that the description of the sets Λ_S we encounter on the way stays within a reasonable complexity. It is, however, indeed possible to make a good selection of primes and to implement the actual computation of Λ_S in a reasonably efficient manner, so that a standard PC (standard as of 2008) can perform the calculations in less than a day. We finally obtain the result we were suspecting from the beginning.

Theorem 4 (Bugeaud, Mignotte, Siksek, Stoll, Tengely).
Let x, y be integers satisfying
$$\binom{y}{2} = \binom{x}{5}.$$
Then $x \in \{0, 1, 2, 3, 4, 5, 6, 7, 15, 19\}$.

A detailed description of the method (explained using the different example equation $y^2 - y = x^5 - x$) can be found in [2].

References

[1] Michael Beck, Eric Pine, Wayne Tarrant, and Kim Yarbrough Jensen, New integer representations as the sum of three cubes. *Mathematics of Computation* **76**(259), 1683–1690 (2007)
[2] Yann Bugeaud, Maurice Mignotte, Samir Siksek, Michael Stoll, and Szabolcs Tengely, Integral points on hyperelliptic curves. *Algebra & Number Theory* **2**(8), 859–885 (2008)
[3] David Hilbert, Mathematische Probleme. Vortrag, gehalten auf dem internationalen Mathematiker-Congress zu Paris 1900 (German). *Nachrichten von der Königlichen Gesellschaft der Wissenschaften zu Göttingen* **1900**, 253–297 (1900)
[4] David Hilbert, Mathematical problems. *Bulletin of the American Mathematical Society* **8**, 437–479 (1902); reprinted: *Bulletin of the American Mathematical Society (New Series)* **37**, 407–436 (2000)
[5] Marc Hindry and Joseph H. Silverman, *Diophantine Geometry. An Introduction.* Graduate Texts in Mathematics, volume 201, Springer, New York (2000)
[6] Yuri V. Matiyasevich, *Hilbert's Tenth Problem.* MIT Press, Cambridge (1993)
[7] Roel J. Stroeker and Benjamin M.M. de Weger, Elliptic binomial Diophantine equations. *Mathematics of Computation* **68**, 1257–1281 (1999)

From Sex to Quadratic Forms

Simon Norton

Abstract. We start with an elementary problem and successively generalize it to reach an important area of mathematics, the theory of quadratic forms. Furthermore we describe a way of calculating the number of essentially different quadratic forms of any discriminant, the class number; this is a concept of great importance, which for example figured in early attempts to prove Fermat's Last Theorem.

1 Introduction

Some time ago I was thinking about a quite simple problem (Problem 1 below) that turned out to lead me via a number of steps to some deep mathematics. I was already fairly familiar with the relevant concepts, but I was led to a new and interesting way of looking at them. The reader is invited to follow my journey.

Problem 1. *A schoolteacher is in charge of some children. She wants to select two of them at random, and notices that it is exactly an even chance (50%) that they are of the same sex. What can be said about how many children of each sex there are?*

We suggest that the reader tries to solve this problem before turning the page.

Simon Norton
Department of Pure Mathematics and Mathematical Statistics, University of Cambridge, Cambridge CB3 0WB, UK. e-mail: S.Norton@dpmms.cam.ac.uk

Let us suppose that there are α boys and β girls. Then the number of (unordered) pairs of children is $\frac{1}{2}(\alpha+\beta)(\alpha+\beta-1)$, while the number of pairs of opposite sex is $\alpha\beta$. So

$$(\alpha+\beta)(\alpha+\beta-1) = 4\alpha\beta \tag{1}$$

which when rearranged gives

$$(\alpha-\beta)^2 = \alpha+\beta. \tag{2}$$

If we write $n = \alpha - \beta \in \mathbb{Z}$, then $\alpha = \frac{n(n+1)}{2}$ and $\beta = \frac{n(n-1)}{2}$. This means that α and β are consecutive triangular numbers, with α being the larger if $n > 0$ and β being the larger if $n < 0$.

The cases where n equals 0 or ± 1 are "degenerate" in the sense that the hypothesis is automatically true because there aren't two children to choose.

Thus we have a complete solution to the problem:

Theorem 1. *The numbers of boys and girls are consecutive triangular numbers. If the solution is to be non-degenerate then the smaller and larger numbers must be at least 1 and 3 respectively. Conversely, any such pair of numbers represents a solution.* □

2 From Sex to Socks

When mathematicians have solved a problem, they tend to look for generalizations. In this case one possible generalization is to assume that the children are of not two but three different sexes. However any mathematician who poses the problem in this form is likely to be seen as being divorced from reality, so let us switch from sex to socks and pose the following problem:

Problem 2. *A man has a selection of socks of three different colours, which he keeps in a bag. Two socks of the same colour may be assumed to form a pair. He finds that if he pulls out two socks at random then there is exactly an even chance that they will form a pair. What can be said about how many socks of each colour he has?*

Let the number of socks of each of the three colours be α, β and γ. We then get the following equations, which correspond to (1) and (2) in Problem 1:

$$(\alpha+\beta+\gamma)(\alpha+\beta+\gamma-1) = 4\beta\gamma + 4\gamma\alpha + 4\alpha\beta \tag{3}$$

$$\alpha^2 + \beta^2 + \gamma^2 - 2\beta\gamma - 2\gamma\alpha - 2\alpha\beta = \alpha+\beta+\gamma \tag{4}$$

We transform the second equation by multiplying by 4 and putting $a = 2\alpha+1$, $b = 2\beta+1$, $c = 2\gamma+1$. This gives us:

$$a^2 + b^2 + c^2 - 2bc - 2ca - 2ab = -3. \tag{5}$$

Remark 1. Our main goal is to classify the solutions of this equation (and its generalizations when -3 is replaced by an arbitrary integer Δ) in (possibly negative) integers, independently of the problem that motivated it. As it turns out, the classification of solutions of this equation, for any value of Δ, provides a classification of binary quadratic forms with discriminant Δ.

Equation (5) can also be written as

$$(a + b - c)^2 = 4ab - 3 \tag{6}$$

so, as the right hand side must be non-negative, a and b (and, similarly, c) must have the same sign. We call a solution of (5) a *positive* or *negative triple* according to the sign of a, b and c. As the equation is invariant under negating a, b and c we may without loss of generality restrict ourselves to positive triples. It also follows from (6) that a and b (and similarly c) must be odd, as otherwise $4ab - 3$ would have remainder 5 when divided by 8, whereas an odd square such as $(a + b - c)^2$ has remainder 1. This therefore proves:

Theorem 2. *There is a $(2,1)$ correspondence between solutions of (5) in integers and solutions of (4) in non-negative integers by changing the signs of a, b and c if they are all negative and then putting $a = 2\alpha + 1$, $b = 2\beta + 1$, $c = 2\gamma + 1$. This in turn gives rise to a non-degenerate solution of Problem 2 (i.e., a solution in which $\alpha + \beta + \gamma$ is at least 2) unless (a, b, c) is $\pm(1, 1, 1)$ or a permutation of $\pm(1, 1, 3)$; though if any of a, b or c is ± 1 not all the three colours will in fact be used.* □

We now study positive triples, and start by using a standard trick to enable us to transform one solution to another. Equation (5) can be rewritten as

$$c^2 - 2(a + b)c + (a - b)^2 + 3 = 0. \tag{7}$$

Consider a particular solution (a, b, c). If we replace c by t we get the equation $t^2 - 2(a + b)t + (a - b)^2 + 3 = 0$, and let us solve this equation for t. As it is a quadratic, it will have two roots, one of them being c. Let us call the other c'. By the remainder theorem the left hand side of the equation is $(t - c)(t - c')$. Equating the coefficients of t, this gives us $c + c' = 2(a + b)$. In other words, if (a, b, c) is a triple, then the other triple with the same values for a and b is $(a, b, 2a + 2b - c)$. Similarly, we may replace a by $2b + 2c - a$ or b by $2a + 2c - b$ in any triple to get another triple. Note that if the triple (a, b, c) is positive, then each of its three transforms must be positive as well, so that $2a + 2b - c$, $2b + 2c - a$, or $2c + 2a - b$ are all positive; also if we repeat any of the three transformations we get back to the original triple.

Transforming a, b or c in this way will simplify the triple (i.e., reduce the value of $|a| + |b| + |c|$) if $a > b + c$, $b > c + a$ or $c > a + b$ respectively. If none of these inequalities hold, so that none of the three transformations simplifies our triple, we call it *reduced*. What reduced triples are there? To find out, we rewrite equation (5) yet again as

$$a(b+c-a) + b(c+a-b) + c(a+b-c) = 3. \qquad (8)$$

By hypothesis, a, b and c are positive. The numbers $b+c-a$, $c+a-b$ and $a+b-c$ are non-negative (because the triple is assumed to be reduced) and odd (because a, b and c are), which means they are positive. So we have to express 3 as the sum of three positive numbers, each of which must therefore be 1, so that the unique reduced triple is $a = b = c = 1$.

Let us consider any positive triple. If it's $(1,1,1)$ then it's reduced. Otherwise we can simplify it by using one of our three transformations. (There's only one way to do this, because at most one of the inequalities $a > b+c$, $b > c+a$, $c > a+b$ can hold.) Unless we have reached $(1,1,1)$ we then repeat the process. By the principle of *infinite descent*, as the numbers keep getting smaller, we must eventually reach a point where no further simplification is possible, which can only be at the unique reduced triple, $(1,1,1)$. We can now reverse the process to get

Theorem 3. *Any positive triple can be obtained from $(a, b, c) = (1, 1, 1)$ by repeatedly applying the three transformations $a \mapsto 2b+2c-a$, $b \mapsto 2c+2a-b$, $c \mapsto 2a + 2b - c$.* □

We note that when we apply these transformations the triple always gets larger: this can be shown by induction, because as we said above at most one of them can reduce the size of the triple, and this can only be the one we used to derive the triple from a smaller one. Analogues of this argument will be used later.

We look at a few triples. If we start with $(a, b, c) = (1, 1, 1)$ and transform b and c alternately, we get $(1, 3, 1)$, $(1, 3, 7)$, $(1, 13, 7)$, $(1, 13, 21)$, and so on. In fact, as $a = 1$ and hence $\alpha = 0$, these are just the triples that correspond to the solutions of Problem 1; $\beta = (b-1)/2$ and $\gamma = (c-1)/2$ are consecutive triangular numbers. If we now transform a in the four triples specified above, we get $(7, 3, 1)$, $(19, 3, 7)$, $(39, 13, 7)$ and $(67, 13, 21)$.

If we study the numbers that appear, one thing stands out: they have no prime factors of the form $3n - 1$. At first glance this seems quite remarkable. We start with a triple $(1, 1, 1)$ and apply some purely additive transformations, and we see a set of numbers none of which has a prime factor of the form $3n - 1$, a multiplicative property.

However, this result is in fact a consequence of (6), because if a was divisible by $p = 3n-1$ then $(a+b-c)^2 +3$ would be too, and it is known that no number of the form $d^2 + 3$ can be divisible by any odd prime of this form.[1] (We already know that p must be odd, as a is.)

---

[1] By Gauss' law of quadratic reciprocity, -3 is the residue of a square modulo an odd prime p if and only if p does not have the form $3n - 1$.

Together with its converse, this gives:

Theorem 4. *A number a occurs in a triple if and only if there is an integer d such that $d^2 + 3$ is divisible by $4a$.*

Proof. If $d^2 + 3$ is divisible by $4a$, we choose $b = (d^2 + 3)/4a$ and $c = a+b-d$. Then $(a+b-c)^2 = d^2 = 4ab - 3$, thus giving a solution to (6), hence (5), in which a appears. The other direction follows directly from (6), as above. □

Remark 2. The condition of Theorem 4 is satisfied, and thus a occurs in a triple, exactly when the odd number a has no prime factor of the form $3n-1$ and is not divisible by 9. We do not prove this here (it uses the fact mentioned in footnote 1 as well as the Chinese Remainder Theorem). Remembering that $a = 2\alpha + 1$, it answers completely the question of what numbers can appear in solutions to Problem 2.

3 From Socks to Triangles

Now we transform our problem into geometric form as a prelude to further generalization.

Consider a triangle with vertices XYZ whose side lengths are $YZ = x = \sqrt{a}$, $ZX = y = \sqrt{b}$ and $XY = z = \sqrt{c}$. Heron's formula states that if $s = \frac{1}{2}(x + y + z)$ is the semi-perimeter of XYZ, then its area is $\sqrt{s(s-x)(s-y)(s-z)}$, or, in terms of a, b and c,

$$\frac{1}{4}\sqrt{-a^2 - b^2 - c^2 + 2bc + 2ca + 2ab}\ .$$

So our triples correspond to triangles whose sides are each the square root of an integer and whose area is $\frac{\sqrt{3}}{4}$ (see (5)).

These properties are still satisfied if we replace the point Z by its image, say Z', under the half turn around the point X. This does not change the side XY, and it takes XZ to XZ' which obviously has the same length. As for the third side, YZ, it is taken to YZ' whose length we may call x'. Also, let us denote the angle at X in the original triangle XYZ by $\hat{X}$; then the corresponding angle in XYZ' will be its supplement $\pi - \hat{X}$.

To show that the areas of XYZ and XYZ' are equal, we note that both triangles share a base XY and that because of the half turn property the corresponding altitudes — the distances of Z and Z' from XY — are equal. To show that x' is the square root of an integer, we apply the cosine rule to the triangles XYZ and XYZ', yielding the equations $x^2 = y^2 + z^2 - 2yz \cos \hat{X}$, $x'^2 = y^2 + z^2 - 2yz \cos(\pi - \hat{X}) = y^2 + z^2 + 2yz \cos \hat{X}$, and therefore $x^2 + x'^2 = 2(y^2 + z^2)$, so since x, y and z are all square roots of integers it follows that x' is. Clearly this is precisely equivalent to the transformation $a \mapsto 2b + 2c - a$ which we obtained previously.

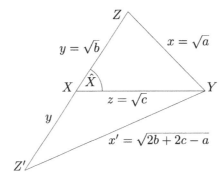

Fig. 1. A triangle XYZ and its image triangle $XZ'Y$ under the half turn operation about X.

This transformation reduces the sum of the squares of the sides of the triangle (which is $|a| + |b| + |c|$) whenever $x' < x$, i.e. (by the cosine rule), whenever $\hat{X}$ is obtuse. A reduced triangle is therefore one which is acute or right angled. By the argument we used to prove Theorem 3 the only reduced triangle with area $\frac{\sqrt{3}}{4}$ is an equilateral triangle with sides 1. The vertices of this triangle determine a hexagonal *lattice*, and all the vertices obtained by repeatedly applying half turn operations also lie on this lattice.

Remark 3. Note that the following interesting statement follows directly from our arguments: Every triangle with area $\frac{\sqrt{3}}{4}$ whose side lengths are square roots of integers can be embedded into a hexagonal lattice of side length 1, i.e., there is a hexagonal lattice of side length 1 such that the vertices of the triangle are lattice points.

It turns out that our operations naturally give rise to a group which we now describe. Consider the set of triangles of area $\frac{\sqrt{3}}{4}$ with labelled vertices and side lengths that are roots of integers, identifying triangles that are translates of each other. Our half turn operations send translated triangles to translated triangles, so they operate on this set.

Let S_0 be the operation that replaces the triangle XYZ with YZX (cyclic permutation of the vertices), and let T_0 be the operation that replaces XYZ with $XZ'Y$ (half-turn of Z around X); we use the image $XZ'Y$ rather than XYZ' in order to preserve orientation.

Since we are ignoring translations, we may represent the triangle XYZ by the two vectors $\overrightarrow{XY}$ and $\overrightarrow{XZ}$. The operation S_0 replaces these vectors by $\overrightarrow{YZ} = \overrightarrow{XZ} - \overrightarrow{XY}$ and $\overrightarrow{YX} = -\overrightarrow{XY}$ respectively, and T_0 replaces them by $\overrightarrow{XZ'} = -\overrightarrow{XZ}$ and $\overrightarrow{XY}$. In other words, S_0 and T_0 take the "vector of vectors" $(\overrightarrow{XY}, \overrightarrow{XZ})$ to

$$(\overrightarrow{XY}, \overrightarrow{XZ}) \begin{pmatrix} -1 & -1 \\ 1 & 0 \end{pmatrix} \quad \text{and} \quad (\overrightarrow{XY}, \overrightarrow{XZ}) \begin{pmatrix} 0 & 1 \\ -1 & 0 \end{pmatrix}$$

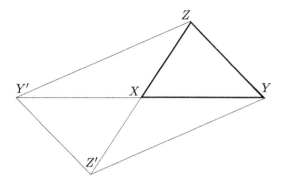

Fig. 2. The operation T_0 sends the triangle XYZ to $XZ'Y$; in turn, it sends $XZ'Y$ to $XY'Z'$, then to XZY' and finally back to XYZ, so applying T_0 four times yields the identity.

so that the two 2×2 matrices above may be thought of as representing S_0 and T_0.

We consider the group $\langle S_0, T_0 \rangle$ generated by S_0 and T_0. As the group composition is compatible with matrix multiplication, we may represent the elements of this group by 2×2 matrices similar to those we used for S_0 and T_0, and it is often useful to think of $\langle S_0, T_0 \rangle$ as a group of 2×2 matrices.

The operations S_0 and T_0 also act as linear transformations on a, b and c. It is easily seen that they postmultiply the vector (a, b, c) by the matrices

$$S = \begin{pmatrix} 0 & 0 & 1 \\ 1 & 0 & 0 \\ 0 & 1 & 0 \end{pmatrix} \quad \text{and} \quad T = \begin{pmatrix} -1 & 0 & 0 \\ 2 & 0 & 1 \\ 2 & 1 & 0 \end{pmatrix}$$

respectively. In fact,

$$(a, b, c)\ T = (-a + 2b + 2c, c, b)$$

and we recognize our formula from above.

In the same way, every transformation $A_0 \in \langle S_0, T_0 \rangle$ has a corresponding operation A acting on triples (a, b, c). A will be generated by S and T in the same way that A_0 is generated by S_0 and T_0, so it acts on these triples as a 3×3 matrix. The mapping sending A_0 to A is thus a homomorphism from $\langle S_0, T_0 \rangle$ to the group of invertible 3×3 matrices.

In fact, this descends to a homomorphism from $\langle S_0, T_0 \rangle / \langle -I \rangle$ to the group of invertible 3×3 matrices[2] because $-I = T_0^2$ is a rotation by $180°$ and thus leaves side lengths invariant. The group $\langle S_0, T_0 \rangle / \langle -I \rangle$ is known as the

---

[2] For readers who are not familiar with group quotients: the notion $\langle S_0, T_0 \rangle / \langle -I \rangle$ denotes the set of all matrices in $\langle S_0, T_0 \rangle$ in which each matrix A is identified with its negative $-A$. This set inherits a natural group structure from $\langle S_0, T_0 \rangle$ and is called the *quotient group* of $\langle S_0, T_0 \rangle$ by $\langle -I \rangle$.

modular group, denoted by Γ, which plays a vital role in many areas of mathematics. We will often write 2Γ for $\langle S_0, T_0 \rangle$ since each element of Γ represents exactly two elements in 2Γ. We also write $\langle S, T \rangle = \Gamma'$, which is isomorphic to Γ, as we will see below.

We will now give an explicit formula for the homomorphism $A_0 \mapsto A$ that we just defined. Note that

$$a = \langle\langle \overrightarrow{YZ}, \overrightarrow{YZ} \rangle\rangle = \langle\langle \overrightarrow{XY}, \overrightarrow{XY} \rangle\rangle - 2\langle\langle \overrightarrow{XY}, \overrightarrow{XZ} \rangle\rangle + \langle\langle \overrightarrow{XZ}, \overrightarrow{XZ} \rangle\rangle,$$

$$b = \langle\langle \overrightarrow{XZ}, \overrightarrow{XZ} \rangle\rangle,$$

$$c = \langle\langle \overrightarrow{XY}, \overrightarrow{XY} \rangle\rangle,$$

where $\langle\langle\,,\,\rangle\rangle$ denotes the inner product. A tedious but not difficult calculation shows that the operation acting as the matrix $\begin{pmatrix} e & f \\ g & h \end{pmatrix}$ on $(\overrightarrow{XY}, \overrightarrow{XZ})$ acts as the following matrix on (a, b, c):

$$\begin{pmatrix} (e-f)(h-g) & -fh & -eg \\ (g-h)(e-f+g-h) & h(f+h) & g(e+g) \\ (e-f)(e-f+g-h) & f(f+h) & e(e+g) \end{pmatrix}.$$

By equating this matrix with the identity it can be seen that the kernel of our group homomorphism is exactly the group generated by $-I$. It follows that the group Γ' generated by S and T, which is the image of the homomorphism, is isomorphic to $\langle S_0, T_0 \rangle / \langle -I \rangle = \Gamma$, as stated above.

Remark 4. By an extension of our arguments it can be shown that any relation between S_0 and T_0 is a consequence of the relations $S_0^3 = I$ and $T_0^2 = -I$, together with the fact that the latter commutes with S_0. This implies that the modular group is the free product of the cyclic groups of orders 3 and 2 generated by the elements of Γ corresponding to S_0 and T_0 respectively.

In the groups Γ' and 2Γ we look at the subgroups $\langle T, STS^{-1}, S^{-1}TS \rangle$ and $\langle T_0, S_0 T_0 S_0^{-1}, S_0^{-1} T_0 S_0 \rangle$. These can easily be seen to have index 3. In the 2×2 representation the generators correspond to the operations that replace Z, X, Y by their images under the half turn about X, Y, Z, and then swap the resulting points with Y, Z, X respectively. In the 3×3 representation they correspond to the operations that transform one of a, b, c as described in Theorem 3, and then swap the other two. We use the notation T_a, T_b, T_c for these operations in the 3×3 representation.

In the literature the modular group is usually defined as the group $\mathrm{PSL}_2(\mathbb{Z})$ of all integral 2×2 matrices with determinant 1, quotienting out $\langle -I \rangle$. Why is this definition equivalent to ours? Well, as S_0 and T_0 have determinant 1, it follows directly that this is true for all elements of $\langle S_0, T_0 \rangle$. On the other hand, let R be a 2×2 matrix with determinant 1. If XYZ is an equilateral triangle with side length 1, then R takes the vector of vectors $(\overrightarrow{XY}, \overrightarrow{XZ})$ to a vector of two vectors spanning a triangle of area $\frac{\sqrt{3}}{4}$ (as R has determinant 1, it can be seen by a straightforward calculation that it preserves areas) which, by the reduction process above, can be obtained from XYZ by applying a transformation from 2Γ. As R coincides with this transformation on the row of vectors $(\overrightarrow{XY}, \overrightarrow{XZ})$, the two transformations have to be equal.

4 From Triangles to Quadratic Forms

Let us again consider the hexagonal lattice generated by an equilateral triangle XYZ of side length 1. A vector between two lattice points will always have the form $m\overrightarrow{XY} + n\overrightarrow{XZ}$ with integers m and n. The length of the vector $m\overrightarrow{XY} + n\overrightarrow{XZ}$ can easily be calculated: by the cosine rule, it equals

$$\sqrt{m^2 + n^2 - 2mn\cos(120°)} = \sqrt{m^2 + mn + n^2}.$$

From the definition of Γ as the group of integral 2×2 matrices of determinant 1 (quotiented out by $\langle -I \rangle$), it follows that a lattice vector can be represented as the first vector of the image of $(\overrightarrow{XY}, \overrightarrow{XZ})$ under an element of 2Γ, and thus occurs as the side of a triangle with area $\frac{\sqrt{3}}{4}$, if and only if it is *primitive*, i.e., none of its submultiples is also a lattice vector. Together with Theorem 4, we get that a positive number a can be written as $m^2 + mn + n^2$ with coprime integers m and n if and only if $4a$ divides $d^2 + 3$ for some integer d. (These numbers are characterized in Remark 2.)

We will now move to a slightly more general situation: Consider a triangular lattice generated by an arbitrary triangle XYZ whose squared side lengths a, b and c are integers. Then for arbitrary integers m and n, the vector $m\overrightarrow{XY} + n\overrightarrow{XZ}$ has squared length

$$\begin{aligned}
|m\overrightarrow{XY} + n\overrightarrow{XZ}|^2 &= \langle\langle m\overrightarrow{XY} + n\overrightarrow{XZ}, m\overrightarrow{XY} + n\overrightarrow{XZ}\rangle\rangle \\
&= m^2\langle\langle \overrightarrow{XY}, \overrightarrow{XY}\rangle\rangle + 2mn\langle\langle \overrightarrow{XY}, \overrightarrow{XZ}\rangle\rangle + n^2\langle\langle \overrightarrow{XZ}, \overrightarrow{XZ}\rangle\rangle \\
&= cm^2 + (-a+b+c)mn + bn^2
\end{aligned}$$

since the inner product in the middle can be written as

$$\frac{1}{2}\left(\langle\langle \overrightarrow{XY}, \overrightarrow{XY}\rangle\rangle + \langle\langle \overrightarrow{XZ}, \overrightarrow{XZ}\rangle\rangle - \langle\langle \overrightarrow{XY} - \overrightarrow{XZ}, \overrightarrow{XY} - \overrightarrow{XZ}\rangle\rangle\right)$$

and the last term in this is $\langle\langle \overrightarrow{YZ}, \overrightarrow{YZ}\rangle\rangle$.

Motivated by this observation, we will now introduce the general concept of *(binary) quadratic forms*. A binary quadratic form is an expression of the form $um^2 + vmn + wn^2$, where u, v and w are fixed integers and m and n are variables. This can be written in matrix form as

$$(m \ \ n) \begin{pmatrix} u & v/2 \\ v/2 & w \end{pmatrix} \begin{pmatrix} m \\ n \end{pmatrix}.$$

In describing a quadratic form $um^2 + vmn + wn^2$, the determinant of the matrix $\begin{pmatrix} u & \frac{v}{2} \\ \frac{v}{2} & w \end{pmatrix}$ plays an important role. Multiplying it by -4, we define the *discriminant* of the quadratic form $um^2 + vmn + wn^2$ as $\Delta = v^2 - 4uw$. As is well known (and not hard to check, using quadratic equations), the quadratic form factorizes into linear factors of the form $sm + tn$ with rational s and t if and only if Δ is a square, and it can in fact be shown that s and t can be chosen as integers.

Note that if we put $a = u + w - v$, $b = u$ and $c = w$, then the discriminant of the quadratic form $um^2 + vmn + wn^2$ becomes

$$v^2 - 4uw = (-a + b + c)^2 - 4bc = a^2 + b^2 + c^2 - 2bc - 2ca - 2ab,$$

i.e., we get exactly the left hand side of equation (5). Thus, *there is a one-to-one correspondence between quadratic forms $um^2 + vmn + wn^2$ with discriminant Δ and solutions (a,b,c) of the equation*

$$a^2 + b^2 + c^2 - 2bc - 2ca - 2ab = \Delta. \tag{5'}$$

Furthermore, if there is a triangular lattice generated by a triangle XYZ which gives rise to the quadratic form $uv^2 + vmn + wn^2$ in the way described above, then the area of this triangle is exactly $\frac{\sqrt{-\Delta}}{4}$, which follows directly from (5') using Heron's formula.

Conversely, we note that if (a, b, c) is a positive triple solving the above equation for negative Δ, then as the left hand side is

$$-(\sqrt{a} + \sqrt{b} + \sqrt{c})(\sqrt{a} + \sqrt{b} - \sqrt{c})(\sqrt{b} + \sqrt{c} - \sqrt{a})(\sqrt{c} + \sqrt{a} - \sqrt{b})$$

it follows that $\sqrt{a}$, $\sqrt{b}$ and $\sqrt{c}$ must satisfy the triangle inequalities, so that there is a triangle with these side lengths which gives rise to the relevant quadratic form.

Summarizing, we see that our classification of integral solutions of (5), and triangles of area $\frac{\sqrt{3}}{4}$ with side lengths which are roots of integers, also classifies quadratic forms of discriminant $\Delta = -3$.

In the next sections, we will do this for all other values of Δ. There are essentially four cases: negative discriminants, zero discriminant, positive square discriminants, and positive non-square discriminants. The first and last of these are the most important, but in the other two cases we can and will give a complete solution.

5 Negative Discriminants

We now pose the following problem, which generalizes the geometric interpretation of Problem 2:

Problem 3. *Find all triangles whose sides are square roots of integers and whose area is $\frac{\sqrt{-\Delta}}{4}$, for any negative integer Δ.*

Our analysis shows that if $\Delta = -3$ we have equation (5), and in general we get equation (5′) above. This is equivalent to equations which similarly generalize (6)–(8), which we call (6′)–(8′). Here, for the reader's convenience, we show equations (6′)–(8′) together:

$$(a+b-c)^2 = 4ab + \Delta, \qquad (6')$$
$$c^2 - 2(a+b)c + (a-b)^2 - \Delta = 0, \qquad (7')$$
$$a(b+c-a) + b(c+a-b) + c(a+b-c) = -\Delta. \qquad (8')$$

Equation (6′) shows that Δ has remainder 0 or 1 when divided by 4. We call numbers of this form *permitted discriminants* (this also applies when Δ is non-negative). Conversely, (6′) has a solution for any Δ of this form: if Δ is divisible by 4 we may put $a = 1$, $b = \frac{-\Delta}{4}$ and $c = b+1$, while if $\Delta - 1$ is divisible by 4 we may put $a = 1$ and $b = c = \frac{1-\Delta}{4}$.

Most of our analysis goes forward with little modification. The set of triples satisfying (5′) is still closed under the half turn operations T_a, T_b, T_c. The definition of reduced triples is unchanged, and equation (8′) can be used to show that there are only finitely many reduced triples, from which all triangles may be obtained by applying a sequence of the operations T_a, T_b and T_c. There may, however, be more than one. Distinct reduced triples are only equivalent under Γ (i.e., there is an operation from Γ' taking the one to the other) if either they are related by a cyclic permutation of (a, b, c), or they are related by an odd permutation and one of a, b and c is the sum of the other two. This is essentially the same as saying that a reduced (a, b, c) is equivalent to (a, c, b) only when two of a, b and c are equal (in other words when the triangle XYZ is isosceles) or when one of them is the sum of the other two (i.e., when the triangle XYZ is right-angled). We call these the *isosceles* and *Pythagorean* cases respectively.

Theorem 3 goes through to say that any positive triple can be obtained from one of the finitely many reduced triples by applying a sequence of the T_i, $i = a, b, c$.

Theorem 4 generalizes to say that a number a occurs in a triple if and only if there is an integer d such that $d^2 - \Delta$ is divisible by $4a$. Our geometric analysis also goes through. As we saw earlier, the area of our triangle XYZ is $\frac{\sqrt{-\Delta}}{4}$, the area of a period parallelogram of the lattice generated by this triangle is $\frac{\sqrt{-\Delta}}{2}$, and the discriminant of the corresponding quadratic form is Δ.

We call a triple *imprimitive* if a, b and c have a common factor, say k. Then the triple $(a/k, b/k, c/k)$ has discriminant Δ/k^2, so, if for some Δ there is no $k > 1$ such that Δ/k^2 is a permitted discriminant, then all triples with discriminant Δ are primitive. This happens when $-\Delta$ is either square-free or 4 times a square-free number of form $4n + 1$ or $4n + 2$. In these cases the number of inequivalent reduced triples is an important function of Δ, called the *class number*.

3: $(1, 1, 1)$
4: $(1, 1, 2)$
7: $(1, 2, 2)$
8: $(1, 2, 3)$
11: $(1, 3, 3)$
12: $(1, 3, 4), (2, 2, 2)$ the first case with distinct reduced triples
15: $(1, 4, 4), (2, 2, 3)$ the first case with distinct primitive
                                           reduced triples
16: $(1, 4, 5), (2, 2, 4)$
19: $(1, 5, 5)$
20: $(1, 5, 6), (2, 3, 3)$
23: $(1, 6, 6), (2, 3, 4), (2, 4, 3)$ the first case where reduced triples appear
                                                  that are neither isosceles nor Pythagorean

Table 1. Reduced triples for the first few values of $-\Delta$. Note that as a cyclic shift of a reduced triple is reduced, we need only show those for which $a \leq b$ and $a \leq c$; furthermore in the isosceles and Pythagorean cases we may also require $b \leq c$.

Remark 5. In these cases the class number is 1 if and only if the ring of so-called *algebraic integers* in the field of rational numbers extended by $\sqrt{\Delta}$ satisfies a property called *unique factorization*. It can be shown that the only cases are $-\Delta = 3, 4, 7, 8, 11, 19, 43, 67$ and 163. This was conjectured in the 19-th century but a complete proof didn't appear till the 1980s. Without going into a definition of algebraic integers, we assert that if $-\Delta$ is divisible by 4 then they are the numbers of the form $m + m'\sqrt{\Delta/4}$ for integers m and m', while if $-\Delta + 1$ is divisible by 4 then they are the numbers of the form $m + m'\sqrt{\Delta}$ where m and m' are either both integers or both halves of odd integers. Nor do we define unique factorization; let us just say that many theorems about integers generalize to rings with this property. The concept of class numbers was motivated by early attempts to prove Fermat's Last Theorem: an attempted proof made an unjustified assumption that the ring of *cyclotomic integers* of order p — integral linear combinations of p-th roots of unity — had unique factorization, and when it was pointed out that this could not be assumed Kummer showed that the equation $x^p + y^p = z^p$ has no solutions in non-zero integers x, y and z whenever the ring of cyclotomic integers of order p has class number not divisible by p.

We digress by noting that the fact that the class number for $\Delta = -163$ is 1 is associated with Euler's famous formula for prime numbers, $x^2 + x + 41$ for $-40 \leq x \leq 39$, and also with the curious fact that $e^{\pi\sqrt{163}}$ is very close to an integer. Both these results also generalize to smaller (negatives of) discriminants with class number 1. In fact, the former result can easily be obtained by our methods: there is exactly one reduced triple for $\Delta = -163$ up to equivalence, which must be $(1, 41, 41)$ as this solves (5') and is reduced. So any other (positive) triple can be obtained from $(1, 41, 41)$ by a sequence of T_i, and by the argument below Theorem 3 we can assume that all these operations make the triple larger. So no number between 2 and 40 can occur in a triple: starting from $(1, 41, 41)$ and applying operations T_i that make the triple larger, the second and third coordinates always remain at least 41, and when one first applies T_a, the first coordinate will change from 1 to $2b + 2c - 1 \geq 41 + 41 - 1 = 81$. By the analogue of Theorem 4, we see that no number between 2 and 40 divides a number of the form $\frac{d^2+163}{4}$.

The rest is straightforward: Assume that $x^2 + x + 41$ is not prime for some $-40 \leq x \leq 39$. Then it has a prime factor p with

$$1 < p \leq \sqrt{x^2 + x + 41} < \sqrt{40^2 + 40 + 41} = 41,$$

and we get a contradiction by noting that

$$x^2 + x + 41 = \frac{(2x+1)^2 + 163}{4}.$$

6 Zero Discriminant

Now we remove the hypothesis that Δ is negative. Of course we can just ask for integer solutions of (5') or for quadratic forms with discriminant Δ, but can we also use the geometric interpretation of Problem 3 without becoming divorced from reality? Yes, if we work in *Lorentz-Minkowski geometry*, the four-dimensional version of which can be defined analytically as the set of points with coordinates (x, y, z, iw) where x, y, z, w are real numbers. This is the geometry associated with Einstein's theory of special relativity. Distances can be either real or pure imaginary numbers, or zero; these correspond respectively to *spacelike, timelike* or *lightlike* vectors. It is possible to find lattices associated with quadratic forms of any permitted discriminant inside this geometry. For example, if we choose the lattices generated by $(0, 1, 0, 0)$ and $(0, 0, 1, i)$ or $(0, 0, 0, i)$ the forms have discriminants 0 and 4 respectively. (However, those who don't want to bother with Lorentz-Minkowski geometry may stick to the algebraic interpretation of equation (5').)

We start with the case when $\Delta = 0$. Equation (6') now reads $(a+b-c)^2 = 4ab$. This means that the product of a and b must be a square, say $a = km^2$

and $b = kn^2$. The equation now yields $c - a - b = \pm 2kmn$, which rearranges to $c = k(m \pm n)^2$. Conversely, $(km^2, kn^2, k(m \pm n)^2)$ is a solution of (6') with $\Delta = 0$, thus giving a complete solution of this case.

This makes sense geometrically, as — assuming for the moment that the triple is positive so that $k > 0$ — it means that one of x, y and z (which, we recall, are the square roots of a, b and c) must be the sum of the other two, and the area of a "triangle" one of whose sides is the sum of the other two is certainly zero. (In fact, points X, Y and Z the right distance apart can be found in a suitable 1-dimensional lattice.)

We summarise these results in:

Theorem 5. *In every (non-negative) triple for $\Delta = 0$, the square roots of a, b and c are integral multiples of the same square root of an integer, and one of the multiples is the sum of the other two.* □

Any number can appear as a value for a. This too makes sense, as the condition in the analogue of Theorem 4 is that there should exist an integer d such that d^2 is divisible by $4a$, which is always true.

7 Positive Discriminants

Let us move to the case where Δ is positive. In this case, for reasons we won't go into, to make the theory of class numbers work, we need to enumerate equivalence classes of quadratic forms not under Γ' but under the larger group Γ'' obtained by adjoining the transformation U that takes

$$(a, b, c) \quad \text{to} \quad (-a, -c, -b).$$

As conjugation by U — the operation replacing X by $U^{-1}XU$ — inverts S and fixes T, it takes Γ' to itself, so Γ' has index 2 in Γ'' and thus the enumerations of equivalence classes under the two groups are very similar.

Problem 4. *Classify quadratic forms of discriminant $\Delta > 0$ under the action of Γ''.*

As when $\Delta < 0$, all quadratic forms are primitive (i.e., their coefficients have no common divisor) if Δ is either square-free or 4 times a square-free number of form $4n + 2$ or $4n + 3$. And in these cases the class number — the number of classes of quadratic forms under the action of Γ'' — is related to the uniqueness of factorization in the ring of algebraic integers in the field of rational numbers extended by $\sqrt{\Delta}$.

When $\Delta > 0$ the concept of a reduced triple doesn't work, as sometimes each of T_a, T_b, T_c makes the triple (a, b, c) larger, but there is a sequence of such transformations that makes it smaller. Instead we work with the set of triples, which we call P, such that (a, b, c) is in P if either at least one of a, b

and c is positive and another is negative, or exactly two of a, b and c are zero. Theorem 6 states the properties of P that make it useful.

Theorem 6. *Every triple is equivalent under Γ' to a triple in P. If two triples each in P are equivalent under Γ', then one can be obtained from the other (up to permutation of the triple) by a sequence of T_i such that at every stage the triple is in P. Finally, there are only finitely many triples in P.*

Proof. If a triple (a, b, c) is not in P, then either a, b, c all have the same sign, or two of them have the same sign and the third is zero. If we can show that in either case one of the T_i makes the triple simpler, the first part of Theorem 6 follows by the infinite descent argument.

This is easy to see in the case when one of a, b and c — say, without loss of generality, a — is zero. Again without loss of generality, we may assume that b and c are positive and that $b \leq c$. If $b = c$ then Δ would be zero, and we are assuming that it is positive. So $b < c$. We may therefore apply T_c, which takes $(0, b, c)$ to $(b, 0, 2b - c)$, which is easily seen to be simpler.

In the other case, a, b and c all have the same sign, which we may assume to be positive. As in the case when Δ is negative we define x, y and z to equal $\sqrt{a}$, $\sqrt{b}$ and $\sqrt{c}$. If x, y and z satisfy the triangle equality (i.e., each is less than the sum of the other two) then a triangle with sides x, y and z can be found in the Euclidean plane and would have area $\frac{\sqrt{-\Delta}}{4}$, which is impossible as Δ is assumed to be positive. So one of x, y and z must be greater than or equal to the sum of the other two — say $z \geq x + y$. Squaring, we get $c \geq a + b + 2xy > a + b$, i.e., $c > a + b$, so, again, replacing c by $2a + 2b - c$ reduces its absolute value.

It is clear that in either of these cases the other two T_i will make the triple larger, and also that the resulting triple will not be in P. That means that if we start with a T_i that takes a triple in P to one that isn't, then any subsequent sequence of T_i can only return to P if the path doubles back on itself. So any sequence of T_i that takes one triple in P to another can be reduced to a sequence such that every intermediate triple is in P, by eliminating subsequences that have no action. That proves the second part of Theorem 6.

To prove the last part, we start with the cases when one of a, b and c is zero, say $a = 0$. Then equation (5') reduces to $(b - c)^2 = \Delta$. If Δ is not a square, this will have no solutions. If it is a square, there will still only be finitely many pairs (b, c) which differ by $\sqrt{\Delta}$ and which do not have the same sign.

This just leaves the cases when two of a, b and c — say b and c — have one sign and the third has the other sign. Then by equation (6') we must decompose Δ into a positive integer $-4ab$ and a non-negative integer $(a + b - c)^2$. There are only finitely many ways of doing this, and for each the factorization of $-4ab$ leads to only finitely many values for a and b, after which the value of $(a + b - c)^2$ leads to at most two values for c. □

8 Orbits of Triples

A general triple has 6 images under the group $\langle S, U \rangle$ (including itself). We may regard these images as essentially the same, as the three numbers, or their negatives, differ only in order. It is convenient to choose a specific representative for each class of 6 triples in P: we choose the one where (a, b, c) has sign pattern $(-, +, +)$, $(-, 0, +)$, $(-, +, 0)$ or $(0, +, 0)$, where in the second and third cases $-a \leq c$ and $-a \leq b$ respectively, and call the set of such triples Q. It is easy to see that every triple in P has exactly one image in Q.

If $(a, b, c) \in P$ contains a zero, then just one of its images under T_a, T_b and T_c will also be in P; otherwise two of them will be. It is sufficient to deal with triples in Q; for the above sign patterns T_a never works, T_b works in the first, second and fourth cases, and T_c in the first and third.

We now define an operation K on a subset of Q. If $(a, b, c) \in Q$ has sign pattern $(-, +, 0)$ then K is undefined. Otherwise we first apply the operation T_b that takes (a, b, c) to $(c, 2a - b + 2c, a)$, which will have one of the sign patterns $(+, +, -)$, $(+, 0, -)$, $(+, -, -)$ or $(0, -, 0)$. We then move this into Q by applying an element of $\langle S, U \rangle$, which will be S^{-1} in the first two cases, U in the third, and US (i.e., U followed by S) in the fourth.

A similar operation can be defined on triples with the first and third sign patterns which applies T_c and then an element of $\langle S, U \rangle$. However it can be shown that this is just the inverse of K (which also takes triples with sign pattern $(0, +, 0)$ to themselves). Note that if K takes a triple to itself so does K^{-1}.

It follows from all of this that all equivalence classes of triples in Q can be expressed by joining each triple to its image under K (and hence also to its image under K^{-1}) and taking connected components of the resulting graph. As there are only finitely many triples in Q, and each is joined to 1 or 2 other triples or just to itself, every connected component is either a (linear) chain, which starts with a triple of sign pattern $(-, 0, +)$ and ends with one of sign pattern $(-, +, 0)$, or a circuit. We now prove:

Theorem 7. *If Δ is not a square then only circuits can happen; if Δ is a square then only chains can happen, except in the trivial case when $a = c = 0$ (which yields a circuit of length 1).*

Proof. The first part is obvious, as the initial triple in a chain has $b = 0$, which means that $\Delta = (a - c)^2$. To prove the second part, it is sufficient to show that if Δ is a square then (a, b, c) is equivalent under Γ to a triple whose first entry is zero, because we showed in Theorem 6 that any equivalence within P can be expressed (up to permutation of the entries) by moving along a chain or circuit.

We move to the geometrical interpretation. As Δ is a square the quadratic form factorizes. Let $sm + tn$ be a factor. Then $t\overrightarrow{XY} - s\overrightarrow{XZ}$ will be a vector of length zero. If this vector is not primitive, take its smallest submultiple

From Sex to Quadratic Forms   37

within the lattice, which will both be primitive and have length zero. But we saw earlier that we can find an element of Γ that turns any primitive lattice vector into a side of our triangle. □

9 Square Discriminants

Note that, as we stated at the end of Section 4, the main interest in this section is that a complete solution is possible, so the reader may choose to skip this section.

If Δ is a square, so that all sequences of K- and K^{-1}-images are chains, then there is a simple description of these chains. To express it, we define a function $L(m,n)$ of two non-negative integers by induction on their size as follows:

(a) If m or $n = 0$ then $L(m,n) = 0$.
(b) If $m \geq n > 0$ then $L(m,n) = L(m-n, n) + 1$.
(c) If $n \geq m > 0$ then $L(m,n) = L(m, n-m) + 1$.

In other words, if we apply the Euclidean Algorithm to m and n, replacing each in turn by its remainder when divided by the other, then $L(m,n)$ is the sum of the resulting partial quotients.

Theorem 8. *Let Δ be a square and let $m \leq n$ be positive integers whose sum squares to Δ. Then we can apply K exactly $L(m,n)$ times to $(-m, 0, n)$ before it stops.*

If m and n are coprime, it stops at $(-q, r, 0)$ where p, q and r are the unique positive integers such that $m + n = q + r = p$, $q \leq r$ and $mq \pm 1$ is divisible by p.

If m and n have greatest common divisor d, then the place where it stops is d times the place where the repeated application of K to $(-\frac{m}{d}, 0, \frac{n}{d})$ stops.

K takes $(0, n, 0)$ to itself.

The above chains, together with the operations S and U, determine all equivalences between triples in P with discriminant Δ.

We start with two lemmas.

Lemma 1. *If $eh - fg = 1$ and e, f, g and h are non-negative, then exactly one of the following three holds:*

A: $e = h = 1$, $f = g = 0$.
B: $e \leq g$ and $f \leq h$.
C: $e \geq g$ and $f \geq h$.

Proof. We leave this as an exercise for the reader. □

Lemma 2. If $\begin{pmatrix} e & f \\ g & h \end{pmatrix} \in 2\Gamma$ has non-negative entries then it can be expressed as a product of a sequence of $T_0 S_0$'s and $T_0^{-1} S_0^{-1}$'s, where the length of the sequence is $\text{Max}(L(e,g), L(f,h))$.

Proof. We start by noting that $T_0 S_0 = \begin{pmatrix} 1 & 0 \\ 1 & 1 \end{pmatrix}$ and $T_0^{-1} S_0^{-1} = \begin{pmatrix} 1 & 1 \\ 0 & 1 \end{pmatrix}$. We now use induction on the size of the entries. By Lemma 1 one of the conditions A–C holds. If A, then the matrix, being the identity, can be expressed as the product of the empty sequence. If B, then

$$\begin{pmatrix} e & f \\ g & h \end{pmatrix} = T_0 S_0 \begin{pmatrix} e & f \\ g-e & h-f \end{pmatrix}$$

where the second matrix on the right hand side is a smaller matrix satisfying the hypothesis of the theorem. So by induction we can express the latter in the required form and then add a $T_0 S_0$ on the left to get an expression for $\begin{pmatrix} e & f \\ g & h \end{pmatrix}$. The argument if condition C holds is similar, using the expression

$$\begin{pmatrix} e & f \\ g & h \end{pmatrix} = T_0^{-1} S_0^{-1} \begin{pmatrix} e-g & f-h \\ g & h \end{pmatrix}.$$

If condition A holds then $L(e,g)$ and $L(f,h)$, hence the Max of these, are zero, which is the length of our sequence. If condition B holds, since $e = 0$ is impossible (because of the condition $eh - fg = 1$), we have $L(e,g) = L(e, g-e) + 1$; while $L(f,h) = L(f, h-f) + 1$ unless $f = 0$, when $L(f,h)$ and $L(f, h-f)$ are both zero. Thus in either case $\text{Max}(L(e,g), L(f,h)) = \text{Max}(L(e, g-e), L(f, h-f)) + 1$, which by induction proves the last part of Lemma 2. The argument for condition C is similar. □

Proof (of Theorem 8). We start by assuming that m and n are coprime. Find the unique g such that $0 < g < n$ and $gm + 1$ is divisible by n, and choose e such that $en - gm = 1$. It is easy to see by induction on (m, n) that $L(m, n) > L(e, g)$. Consider the sequence of $L(m, n)$ matrices of the form $T_0 S_0$ or $T_0^{-1} S_0^{-1}$ whose product is $\begin{pmatrix} e & m \\ g & n \end{pmatrix}$. In the 3-dimensional representation, the corresponding matrix is

$$\begin{pmatrix} (e-m)(n-g) & -mn & -eg \\ (g-n)(e-m+g-n) & n(m+n) & g(e+g) \\ (e-m)(e-m+g-n) & m(m+n) & e(e+g) \end{pmatrix}.$$

Using the equation $en - gm = 1$, we find that this takes $(0, m, -n)$ to $(m - e + n - g, 0, -e - g)$. We can apply elements of $\langle S, U \rangle$ before and after to get something that takes $(-m, 0, n)$ to whichever of $(-e-g, m-e+n-g, 0)$ and $(e-m+g-n, e+g, 0)$ — say $(-q, r, 0)$ — is in Q.

Each $T_0 S_0$ and $T_0^{-1} S_0^{-1}$ will correspond to an application of K in a chain that starts with $(-m, 0, n)$ and ends with $(-q, r, 0)$. So we need exactly $L(m, n)$ applications. It is clear that $q + r = m + n$, so all that remains to be checked is that $mq \pm 1$ is divisible by $m + n$. But if $q = e + g$ then $mq + 1 = m(e + g) + 1 = e(m + n)$, and if $q = m - e + n - g$ then $mq - 1 = m(m + n) - m(e + g) - 1 = (m - e)(m + n)$.

If m and n have greatest common divisor $d > 1$, we just apply the argument to $(-\frac{m}{d}, 0, \frac{n}{d})$ and multiply by d. And the case of $(0, n, 0)$ is even more obvious. Finally, the last paragraph of Theorem 8 is an immediate consequence of Theorem 7. □

This leaves only the most interesting case, where the discriminant is positive but not a square.

10 Positive Non-Square Discriminants

We turn to the problem of counting the number of circuits when Δ is a positive non-square. In this case the only sign pattern that occurs in Q is $(-, +, +)$.

It can be seen that if one interchanges b and c in each triple of a circuit one gets either the same or a different circuit in reverse cyclic order. Let us therefore call $(-a, c, b)$ the *reflection* of $(-a, b, c)$. Reflection is actually the product of the operation U with negation, and what we are saying is that conjugating by it inverts the operation K. Now certain triples play a special role under reflection:

1. The triple $(-a, b, b)$ is its own reflection. This gives a (Lorentz-Minkowski) isosceles triangle, so as in the case $\Delta < 0$ we call this the *isosceles* case (I). In this case $\Delta = a(a + 4b)$.
2. The triple $(-a, b, a)$ is taken to its own reflection by K. As the squared length of one of the sides of the corresponding triangle is the negative of the squared length of another, let's call this the *anti-isosceles* case (A). In this case $\Delta = 4a^2 + b^2$.
3. The triple $(-a, b - a, b)$ is also taken to its own reflection by K. As the squared length of one of the sides is the sum of those of the other two, as before we call this the *Pythagorean* case (P). In this case $\Delta = 4ab$.

One can show that any triple that is, or is taken by K to, its reflection must be of one of these three types. It follows that if a circuit with length at least 2 is its own reflection then it must contain exactly two triples of one of these types. So if every circuit is self-reflecting and of length at least 2, then the class number will be half the number of triples of types A, I or P.

The reader may find it an interesting exercise to enumerate all the circuits for various values of Δ (either by hand or by writing a computer program), study their behaviour, and try to prove any results that turn up. I wrote a computer program covering all $\Delta < 500$ and found (and then proved) several interesting results. Here are some of them:

- A primitive circuit of length 1 only occurs when $\Delta = 5$. The relevant triple is $(-1, 1, 1)$. As this is both isosceles and anti-isosceles, and the only primitive triple (for any positive non-square value of Δ) that satisfies more than one of the above conditions to lie in A, I or P, it may be thought of as a "degenerate" example of the result that every self-reflecting circuit has 2 triples that lie in A, I or P.
- A primitive circuit of length 2 occurs only when $\Delta = 8$, and has triples $(-1, 1, 2)$ (which is Pythagorean) and $(-1, 2, 1)$ (which is anti-isosceles).
- The number of primitive triples of each of the types A, I or P is always either zero or a power of 2 (including 1), the same in each case where it is non-zero. It will be non-zero for at least one of I and P, and zero in at least one other case. It follows from this that the number of self-reflecting circuits — and therefore the class number, if all circuits are self-reflecting and all triples are primitive — is always a power of 2.
- In the above, the types that occur, and what the power of 2 is, can be deduced from the number of odd primes dividing Δ, whether any of them have the form $4m - 1$, the power of 2 dividing Δ, and (if this is 4) whether $\Delta/4$ is of form $4m + 1$ or $4m - 1$.
- Self-reflecting circuits can in principle be of six types: AA (i.e., containing two anti-isosceles triples), AI, AP, II, IP, and PP. For each of the last five, there are values of Δ where the answer to the previous result forces the existence of such circuits (e.g. 13, 8, 21, 12 and 24 respectively). Though not forced in this way, the type AA also occurs, but the smallest value of Δ is 136.
- When Δ is small, then all circuits are self-reflecting. The smallest counterexample is when $\Delta = 145$, which has two self-reflecting circuits plus a pair related by reflection, giving a class number of 4. An example of a triple of discriminant 145 that generates a circuit that isn't self-reflecting is $(-8, 2, 3)$.
- The next example of a non self-reflecting circuit occurs when $\Delta = 148$, generated for example by $(-7, 3, 4)$. As there is a unique primitive self-reflecting circuit, the total number of primitive circuits is 3, and this is the first case when this number not a power of 2.
- The next example after that is when $\Delta = 229$, generated for example by $(-9, 3, 5)$. Again there is a unique self-reflecting circuit, and as 229 is square-free this gives the first example of a quadratic field whose class number, 3, is not a power of 2.
- The first Δ with more than 2 primitive self-reflecting circuits is 480.

11 Conclusion

Our journey, which started with an elementary problem, has taken us through several different branches of mathematics, thus illustrating the unity of the subject. We have introduced two important concepts: the modular group and class numbers for rings of algebraic integers in algebraic number fields. Furthermore we have described a method of calculating class numbers of quadratic number fields.

We should note that there is a different, more classical approach to calculating the class numbers of quadratic number fields. It is less explicit than our approach, but leads to a closed formula for the class number; see for instance [2].

We hope that this article will stimulate readers to find out more for themselves; we added a few suggestions for further reading below.

Further Reading

[1] Godfrey H. Hardy and Edward M. Wright, *An Introduction to the Theory of Numbers, sixth edition.* Oxford University Press, Oxford (2008)
This is a "classical" textbook on Number Theory; it provides background and details to the topics of the article. Particularly relevant are Chapter XIV (quadratic fields (1)) and Chapter XV (quadratic fields (2)), as well as Chapter VI (Fermat's theorem and consequences, including quadratic reciprocity).

[2] Zenon I. Borevich and Igor R. Shafarevich, *Number Theory.* Academic Press, New York/London (1966)
A well-written textbook on number theory, more advanced than [1]. Includes a classical closed formula for the class number (Chapter 5.4).

[3] Duncan A. Buell, *Binary Quadratic Forms. Classical Theory and Modern Computations.* Springer, New York (1989)
This is a general textbook on quadratic forms, including an extensive collection of tables extending those given in our text.

Small Divisors: Number Theory in Dynamical Systems

Jean-Christophe Yoccoz

Abstract. We discuss dynamical systems with two or more moving particles, such as two planets orbiting around the sun. If the ratio of their rotation periods, say α, is rational, then the planets are in resonance, and the mutual interaction will make the dynamics unstable. If the period ratio α is irrational, it can be approximated arbitrarily well by rational numbers, and the stability depends on how good this approximation is in terms of the sizes of numerators and denominators. We discuss this in a mathematical model case that can be analyzed completely, the setting of iteration of quadratic polynomials $z \mapsto e^{2\pi i \alpha} z + z^2$, and show how this leads to questions of Diophantine approximation within number theory. Finally, we briefly mention the situation of more than two planets.

1 Planetary Systems

Celestial mechanics is the study of the motion of celestial bodies under Newton's law of gravitation. This law stipulates that the attractive force between any two bodies (which we assume to have negligible size) is proportional to their masses and inversely proportional to the square of their distance. The acceleration of each body is proportional to the resultant of the forces to which it is submitted. In mathematical terms, the motion of N bodies in gravitational interaction is therefore the solution of a system of differential equations of second order.

When there are only 2 bodies, this system can actually be solved explicitly and leads to the famous Kepler laws, which were discovered experimentally long before Newton's law was established: the bodies either escape to infinity

---

Jean-Christophe Yoccoz
Collège de France, 3 rue d'Ulm, 75231 Paris Cédex 05, France.
e-mail: jean-c.yoccoz@college-de-france.fr

(the uninteresting case) or they move periodically along homothetic elliptic orbits.

When there are at least 3 bodies, the system of differential equations becomes fantastically complicated and remains largely mysterious even today. Poincaré showed at the end of the 19-th century that in some appropriate sense, one cannot write the solutions of the system through explicit formulas (somewhat like Galois's assertion, several decades earlier, about the impossibility to solve by radicals the general polynomial equation of degree 5 or larger). Poincaré then looked for other methods in order to study the solutions, founding the modern theory of dynamical systems [8].

Planetary systems constitute a particularly interesting special case of the general N-body problem. One of the bodies (the sun) is assumed to be much heavier than the others (the planets). Therefore, in a first approximation, one can forget about the gravitational interaction between the planets. Each planet will then, independently of the others, move periodically along an ellipsis with the sun as focus. When the motions of all the planets are considered together, the motion is no longer periodic unless the periods of the planets are all commensurate (i.e. all periods have a common multiple): such a superposition of periodic motions (with not necessarily commensurate periods) is called *quasiperiodic*.

A major question is to understand how much this picture changes when one takes into account the mutual gravitational attraction between the planets. In the short or medium term (a few revolutions around the sun), the effect will not be very important because the perturbation is so much smaller than the attractive force of the sun. But in the long term, the effect is quite significant, at least when some periods are close to being commensurate.[1] For instance, the period of Jupiter is close to 2/5 of the period of Saturn, and for the orbits of these two planets this produces deviations from the Keplerian solutions that have been documented by astronomers several centuries ago.

The question of stability of quasiperiodic motions under small perturbations has been one of the major areas of research in dynamical systems theory for one century. Negative results appeared in the first decades of the 20-th century. Then in 1942 Siegel achieved the first breakthrough in a setting which is described below. In the setting of mechanics which is appropriate for planetary systems, a number of results have been obtained since the 1950s; these are collectively known as KAM theory after Kolmogorov, Arnol'd, and Moser who were the pioneers in this line of research. A very good survey is [1].

---

[1] Of course, every real number is arbitrarily close to rational numbers; what matters is how well a real number (the ratio of the periods) can be approximated by rationals in relation to the sizes of numerators and denominators. This will be a key theme in our discussion below.

2 Complex Quadratic Polynomials and Linearization

In this section, we will consider sequences $(z_n)_{n\geq 0}$ of complex numbers which are defined by their initial term z_0 and some recurrence relation $z_{n+1} = f(z_n)$. The map f is fixed and we want to understand the behaviour of the sequence $(z_n)_{n\geq 0}$ as the integer n (that should be thought of as time) goes to ∞. This in general requires different tools depending on the nature of the transformation f; we will only consider examples related to the stability of quasiperiodic motions. A good general reference on the topic of this section, and more generally on complex dynamics, is [7].

The reference example of a pure unperturbed quasiperiodic motion is given by

$$z_{n+1} = \lambda z_n ,$$

i.e. $f(z) = \lambda z$. Here λ is a fixed complex number of absolute value 1; such a number can be written uniquely as $\lambda = \exp(2\pi i \alpha)$ where α is a real number in $[0,1)$. Geometrically, z_{n+1} is obtained from z_n by a rotation of angle $2\pi\alpha$ centered at the origin of the complex plane. The sequence of points $z_0, z_1, z_2, \ldots, z_n, \ldots$ is called the *orbit* of the initial point z_0. (The orbit points z_n can be thought of as the positions at time n of a small planet that orbits around the origin. Here the unit of time is chosen arbitrarily; it will be specified below.) This is the unperturbed system. For this very simple example, we can find an explicit formula for the full sequence:

$$z_n = \lambda^n z_0 = \exp(2\pi i n \alpha) z_0 .$$

We thus have to distinguish two cases:

- α is a rational number $\frac{p}{q}$ (with coprime p and q). In this case, we have $\lambda^q = 1$, hence $z_{n+q} = z_n$ for all $n \geq 0$ and the sequence (z_n) is periodic of period q.
- α is an irrational number. Except in the trivial case $z_0 = 0$, the z_n are all distinct and lie on the circle centered at the origin of radius $|z_0|$. It is not difficult to show that the sequence z_n is actually dense on this circle: for every point z on the circle and every $\delta > 0$, there exist infinitely many z_n whose distance to z is smaller than δ.

We will now consider a very specific perturbation of the previous example where the recurrence relation is

$$z_{n+1} = \lambda z_n + z_n^2 ,$$

i.e., $f(z)$ is the complex quadratic polynomial $\lambda z + z^2$. We will assume that the initial value z_0 is small (in absolute value); for small z, the perturbation (the quadratic term z^2) is much smaller than the linear term λz and the new example is indeed a small perturbation of the previous one.

This quadratic map is particularly important because it is the simplest non-trivial perturbation of the unperturbed system $z \mapsto \lambda z$. It can model the dynamics of two weakly interacting planets as follows. Suppose the ratio of the orbit periods of planets 2 and 1 is α; then during one period of planet 2, planet 1 moves through an angle $2\pi\alpha$. Choosing the unit of time as the period of planet 2, the motion of planet 1 is described by the map $z_{n+1} = \lambda z_n$ with $\lambda = e^{2\pi i \alpha}$. The term z_n^2 models the combined effect of the perturbation from planet 2 on planet 1 during one orbit period of planet 2. (To illustrate that this perturbation should be small, one could write $z_{n+1} = \lambda z_n + \varepsilon z_n^2$; but in coordinates $w_n = \varepsilon z_n$, one obtains again $w_{n+1} = \lambda w_n + w_n^2$.)

The case of rational α is quite interesting [7, Sec. 10], but we will limit ourselves to the following simple remark. Assume that $\alpha = 0$, so that $z_{n+1} = z_n + z_n^2$, and that z_0 is real and close to 0. Then the sequence (z_n) is converging to 0 if $z_0 < 0$, and increasing to $+\infty$ if $z_0 > 0$. So the behaviour is completely different from the unperturbed case $z_{n+1} = z_n$.

Now we generalize and ask the natural question whether the following property holds for arbitrary α:

(Bdd)   The sequence (z_n) is bounded when $z_0 \in \mathbb{C}$ is close enough to the origin.

We have just seen that this is not the case for $\alpha = 0$, and the same holds actually for any rational α [7, Lemma 11.1]. For the unperturbed linear example, the answer obviously is yes for all $\alpha \in [0, 1)$, rational or not.

Let us consider the following (apparently) much stronger property called *linearizability*:

(Lin)   In a neighbourhood of the origin there exists a change of variables $z = h(y)$, defined by a bijective complex differentiable[2] map $h(y)$ with $h(0) = 0$, such that after setting $y_n = h^{-1}(z_n)$ the recurrence relation $z_{n+1} = \lambda z_n + z_n^2$ is transformed into $y_{n+1} = \lambda y_n$.

In other terms, the change of variables h must turn the perturbed linear map $f: z \mapsto \lambda z + z^2$ into the exactly linear map $y \mapsto \lambda y$ (it *linearizes* f). This means that h must satisfy the functional equation $h^{-1} \circ f \circ h(y) = \lambda y$, or equivalently

(FE) $$\lambda h(y) + h(y)^2 = h(\lambda y) ,$$

for y close to the origin (see Figure 1).

---

[2] According to a basic result in complex analysis, a function h is complex differentiable in a neighbourhood of the origin if and only if it can be expressed as a power series $h(z) = \sum_{\ell \geq 0} h_\ell z^\ell$, where the growth of the $|h_\ell|$ is modest enough so that this converges for all z with $|z| < r$, for some $r > 0$. Such functions h are also called *holomorphic*. A holomorphic map is invertible in a neighbourhood of the origin if and only if $h_1 \neq 0$; in this case the inverse is holomorphic as well.

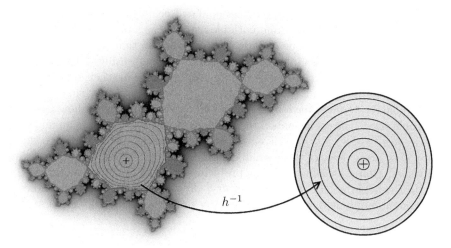

Fig. 1. Left: The dynamics of a polynomial $f(z) = \lambda z + z^2$ with $|\lambda| = 1$ that satisfies condition (Bdd). (Here $\lambda = e^{2\pi i \alpha}$ with $\alpha = (\sqrt{5} - 1)/2$, the "golden mean".) The set of points $z_0 \in \mathbb{C}$ for which the sequence z_n is bounded is coloured orange; its boundary is called the *Julia set*. Condition (Bdd) means that the origin (marked by a '+') has an orange neighbourhood. The largest such open neighbourhood is called a *Siegel disk*. There is a change of variables h^{-1} from the Siegel disk to a round disk D (right) that transports the dynamics of f to multiplication by λ, i.e., to a rigid rotation by the angle α: in other words, $h^{-1} \circ f \circ h(y) = \lambda y$ for all $y \in D$, so condition (Lin) holds. For every point $y_0 \in D$ other than 0, the points $y_n = \lambda^n y_0$ lie densely on a circle; some such circles are drawn in, together with their image circles under h in the Siegel disk. (Left picture courtesy of Arnaud Chéritat.)

Property (Lin) means that the behaviour of the sequences (z_n) is deformed, but not qualitatively changed, by the introduction of the quadratic term z_n^2 in the recurrence relation: therefore, (Lin) implies (Bdd).

On the other hand, if property (Lin) is not satisfied, property (Bdd) is also not satisfied: there always exist arbitrarily small initial values z_0 such that the sequence (z_n) is unbounded. We cannot prove this here; see [7, Lemma 11.1]. Hence, Properties (Bdd) and (Lin) are in fact always equivalent (for holomorphic maps).

A related important condition is *stability* of the fixed point at the origin: this means that for every $\varepsilon > 0$ there is a $\delta > 0$ so that for any point z with distance at most δ from the origin, the entire orbit has distance at most ε from the origin. Obviously, (Lin) implies stability, and stability implies (Bdd), so all three conditions are equivalent for holomorphic maps.

It is actually quite easy to see that for rational α, property (Lin) can never be satisfied for any polynomial f of degree 2 (or higher). If, say, $\alpha = p/q$, then the q-th iterate of the linear map equals the identity. If (Lin) were satisfied, then the origin would have to have a neighbourhood in which the q-th iterate of f was equal to the identity; but the q-th iterate is a polynomial of degree 2^q, and hence has only finitely many fixed points.

We will from now on consider irrational α. Recall the function h from condition (Lin). As a complex differentiable map, it has a power series expansion $h(y) = y + \sum_{\ell \geq 2} h_\ell y^\ell$ (one can always normalize so that $h_1 = 1$). The coefficients h_ℓ can, in principle, be calculated recursively using (FE). As ℓ gets large, one obtains ever more complicated formulas for h_ℓ; their denominators contain factors of the form $\lambda^j - 1$. These factors can be very small (for any irrational α, we must have $\inf_{n \geq 1} |\lambda^n - 1| = 0$). Depending on how fast they become small, the coefficients h_ℓ can grow very rapidly, and this can make the series $h(y) = y + \sum_{\ell \geq 2} h_\ell y^\ell$ divergent even for arbitrarily small nonzero values of y. The occurrence of these small denominators gave rise to the term "small divisors" that was coined for this problem and others of the same nature.

In the beginning of the 20-th century, Cremer [3] constructed examples of irrational numbers α which do not satisfy property (Lin). Observe first that

$$\begin{aligned} z_1 &= \lambda z_0 + z_0^2, \\ z_2 &= \lambda^2 z_0 + (\lambda + \lambda^2) z_0^2 + 2\lambda z_0^3 + z_0^4, \\ z_3 &= \lambda^3 z_0 + \cdots + z_0^8, \end{aligned}$$

and more generally

$$z_n = \lambda^n z_0 + \cdots + z_0^{2^n} =: P_{n,\lambda}(z_0).$$

If z_0^* is a solution of $P_{n,\lambda}(z_0) - z_0 = 0$, the sequence (z_n) with initial value $z_0 = z_0^*$ is periodic of period n. By Viète's Law, the product of the $2^n - 1$ nonzero solutions equals $1 - \lambda^n$. Therefore, there exists such a solution satisfying

$$|z_0^*| \leq |\lambda^n - 1|^{\frac{1}{2^n - 1}}.$$

Assume that λ satisfies

(Cr) $$\inf_{n \geq 1} |\lambda^n - 1|^{\frac{1}{2^n - 1}} = 0.$$

Then we conclude that there are periodic sequences (z_n) starting arbitrarily close to 0 (but not at 0). But then property (Lin) is not satisfied, because the unperturbed reference example does not have such periodic sequences.

It remains to see that there are irrational numbers α that satisfy (Cr). To define such a number, let $b_0 = 2$, $b_{k+1} = b_k^{2^{b_k}}$ for $k \geq 0$, and $\alpha = \sum_{k \geq 0} b_k^{-1}$. This is clearly an irrational number (as can be seen most easily when writing it in base 2). We evaluate $|\lambda^n - 1|$ for $n = b_k$ and get

$$\begin{aligned} |\lambda^{b_k} - 1| &= \left| \exp\left(2\pi i b_k \sum_{\ell \geq 0} b_\ell^{-1}\right) - 1 \right| = \left| \exp\left(2\pi i b_k \sum_{\ell > k} b_\ell^{-1}\right) - 1 \right| \\ &\approx 2\pi b_k / b_{k+1} = 2\pi b_k^{-(2^{b_k} - 1)}, \end{aligned}$$

and thus $|\lambda^{b_k} - 1|^{\frac{1}{2^{b_k} - 1}} \approx 1/b_k$.

This concludes our discussions on angles α for which condition (Lin) is not satisfied, and we turn our attention to the converse. In 1942, Siegel [9] proved the following remarkable result:

Theorem 1 (Siegel). *If λ satisfies the Diophantine Condition*

$$(\text{DC})_{\gamma,\tau} \qquad\qquad |\lambda^n - 1| \geq \frac{\gamma}{n^{1+\tau}}$$

for some constants $\gamma > 0$, $\tau \geq 0$ and for all $n > 0$, then property (Lin) holds.

Siegel's result holds for perturbations of the reference linear example which are much more general than the quadratic one that we have considered. More precisely, it holds for recurrence relations of the form

$$z_{n+1} = \lambda z_n + g(z_n),$$

provided g is complex differentiable in a neighbourhood of the origin and satisfies $g(0) = 0$ and $g'(0) = 0$.

Conditions such as (Cr) and $(\text{DC})_{\gamma,\tau}$ are related to the approximation of irrational numbers by rational numbers, a subject discussed in the next section. For now, we define a fundamental concept and relate it to condition $(\text{DC})_{\gamma,\tau}$ above.

Definition (Diophantine Numbers). An irrational number α is called *Diophantine of exponent τ* if there exists some $\gamma > 0$ such that all $p, q \in \mathbb{Z}$ with $q > 0$ satisfy $|\alpha - p/q| > \gamma/q^\tau$.

It is easy to see that a number α is Diophantine of exponent $2 + \tau$ if and only if $\lambda = e^{2\pi i \alpha}$ satisfies Condition $(\text{DC})_{\gamma,\tau}$ for some $\gamma > 0$.

3 Diophantine Approximation

Given an irrational number α and any $\varepsilon > 0$, there exists a rational number $\frac{p}{q}$ such that $|\alpha - \frac{p}{q}| < \varepsilon$. But as ε becomes small, q (and $p \approx \alpha q$) must become large. How fast in terms of ε?

The *continued fraction algorithm* produces, for every irrational number α, a sequence of rational numbers (p_k/q_k) called the *convergents* of α which are, in a sense explained below, the best rational approximations of α. The algorithm also analyses the quality of these approximations. Background and more information on continued fractions and Diophantine approximation can be found in [4, Secs. X–XI], as well as [7, Sec. 11].

We denote by $[x]$ the integral part of a real number x and by $\{x\}$ its fractional part. Given an irrational number α, define $a_0 = [\alpha]$, $\alpha_1 = \{\alpha\}$, and then $a_k = [\alpha_k^{-1}]$, $\alpha_{k+1} = \{\alpha_k^{-1}\}$ for $k \geq 1$. Thus, one obtains recursively

(CF) $$\alpha = a_0 + \cfrac{1}{a_1 + \cfrac{1}{a_2 + \cfrac{1}{a_3 + \cdots}}}.$$

For $k \geq 0$, define the k-th *convergent* of α as

$$\frac{p_k}{q_k} = a_0 + \cfrac{1}{a_1 + \cfrac{1}{\ddots + \cfrac{1}{a_k}}}$$

(in lowest terms). The sequences of integers (p_k), (q_k) satisfy the following recurrence relation [4, Sec. 10.2]:

$$p_k = a_k p_{k-1} + p_{k-2}, \quad q_k = a_k q_{k-1} + q_{k-2}$$

starting with $p_{-2} = q_{-1} = 0$, $p_{-1} = q_{-2} = 1$. For instance, for the *golden mean* $\alpha = \frac{\sqrt{5}+1}{2}$, all a_k are equal to 1 and $(p_k = q_{k+1})$ is the Fibonacci sequence.

Conversely, for *any* sequence (a_k) of integers with $a_k \geq 1$ for $k \geq 1$, the formula (CF) defines a unique irrational number α.

The convergents are the best rational approximations of α in the following sense: let $k \geq 0$ and let p, q be integers with $0 < q < q_{k+1}$; if one has

$$|q\alpha - p| \leq |q_k \alpha - p_k|,$$

then $q = q_k$ and $p = p_k$ [4, Sec. 10.15].

Concerning the quality of the approximation by the convergents, one has for all $k \geq 0$ the following estimates:[3]

$$\frac{1}{(a_{k+1}+2)q_k} \leq \frac{1}{q_{k+1}+q_k} < |q_k \alpha - p_k| < \frac{1}{q_{k+1}} \leq \frac{1}{a_{k+1}q_k}.$$

---

[3] Since the central inequalities are not as easily found in the introductory literature, we justify them here. We use the fact that the convergents p_k/q_k converge alternatingly to α, in the sense that $p_{2k}/q_{2k} < p_{2k+2}/q_{2k+2} < \cdots < \alpha < \cdots < p_{2k+3}/q_{2k+3} < p_{2k+1}/q_{2k+1}$ for all k. Moreover, $p_{k+1}q_k - q_{k+1}p_k = (-1)^k$, which is easily proved by induction using the recursive relations for p_k and q_k. This gives $\left|\alpha - \frac{p_k}{q_k}\right| < \left|\frac{p_{k+1}}{q_{k+1}} - \frac{p_k}{q_k}\right| = \frac{1}{q_k q_{k+1}}$.

For the second inequality, we repeatedly use the fact that $\frac{a}{c} < \frac{b}{d}$ for $a, b, c, d > 0$ implies $\frac{a}{c} < \frac{a+b}{c+d} < \frac{b}{d}$. We thus get (for even k; the other case is analogous)

$$\frac{p_k}{q_k} < \frac{p_k + p_{k+1}}{q_k + q_{k+1}} \leq \frac{p_k + a_{k+2}p_{k+1}}{q_k + a_{k+2}q_{k+1}} = \frac{p_{k+2}}{q_{k+2}} < \alpha < \frac{p_{k+1}}{q_{k+1}}$$

and thus

$$\left|\alpha - \frac{p_k}{q_k}\right| > \left|\frac{p_k + p_{k+1}}{q_k + q_{k+1}} - \frac{p_k}{q_k}\right| = \left|\frac{p_{k+1}q_k - p_k q_{k+1}}{q_k(q_k + q_{k+1})}\right| = \frac{1}{q_k(q_k + q_{k+1})}.$$

Small Divisors: Number Theory in Dynamical Systems 51

Therefore, large a_{k+1} correspond to especially good rational approximations of α. The golden mean is thus the irrational number with the worst rational approximations, and hence the best candidate for a rotation angle α satisfying condition (Lin) and thus stability.

From the last inequalities above, together with the recursion formula $q_{k+1} = a_{k+1} q_k + q_{k-1}$, it is easy to see that a number α is Diophantine of exponent $2 + \tau$ if and only if

$$q_{k+1} = O(q_k^{1+\tau}) \quad \text{or equivalently} \quad a_{k+1} = O(q_k^\tau)$$

(this just means that the sequences $q_{k+1}/q_k^{1+\tau}$ and a_{k+1}/q_k^τ are bounded). For instance, the golden mean is Diophantine of exponent 2. Note that the golden mean satisfies the equation $\alpha^2 = \alpha + 1$. More generally, for any irrational number α that is the root of a second-degree polynomial with integer coefficients, the sequence (a_k) becomes periodic for large k, so this sequence is bounded, and α is Diophantine of exponent 2 [4, Sec. 10.9].

Many more Diophantine numbers are provided by the following theorem, the proof of which is rather elementary (see for instance [7, Theorem 11.6] or [4, Sec. 11.7]).

Theorem 2 (Liouville). *If α is an irrational number which is the root of a polynomial of degree $d \geq 2$ with integer coefficients, then α is Diophantine of exponent d.*

A much deeper and more difficult theorem in this direction is Roth's theorem.

Theorem 3 (Roth). *If α is an irrational number which is the root of a polynomial (of any degree) with integer coefficients, then α is Diophantine of exponent $2 + \tau$ for any $\tau > 0$.*

Assume that you choose at random a number $\alpha \in [0,1)$, by choosing successively and independently the digits in its decimal expansion with equal probability. Then α will be irrational almost surely (i.e. with probability 1) and the corresponding sequences (a_k), (q_k) will almost surely satisfy the following properties:

- the sequence $\left(\dfrac{a_k}{k \log k}\right)_{k \geq 2}$ is unbounded;
- for any $\varepsilon > 0$, the sequence $\left(\dfrac{a_k}{k(\log k)^{1+\varepsilon}}\right)_{k \geq 2}$ is bounded;
- the sequence $\left(\dfrac{1}{k} \log q_k\right)$ converges to $\dfrac{\pi^2}{12 \log 2}$.

This implies that a random α is almost surely Diophantine of exponent $2 + \tau$ for every $\tau > 0$. The corresponding numbers $\lambda = e^{2\pi i \alpha}$ thus satisfy the conditions of Siegel's theorem so that the corresponding recurrence relation $z_{n+1} = \lambda z_n + g(z_n)$ satisfies condition (Lin).

Remark. The first two assertions are special cases of a more general theorem due to Khinchin; see [6, Section II] or [5, Theorem 30]. The third assertion is essentially Lochs' theorem: it easily implies that almost surely the number of valid decimal places of α in p_k/q_k, divided by k, tends to $\pi^2/6 \log 2 \log 10 \approx 1.0306\ldots$ for random numbers α (in other words, each extra term in the continued fraction expansion of α gains just over one more decimal digit, on average).

4 Further Results and Open Questions

Let α be an irrational number and $\lambda = \exp(2\pi i \alpha)$. We have seen above that if the convergents (p_k/q_k) of α satisfy

(DC) $$q_{k+1} = O(q_k^{1+\tau})$$

for some $\tau \geq 0$ then λ satisfies (DC)$_{\gamma,\tau}$ for some $\gamma > 0$ and thus, by Siegel's theorem, the sequences defined by $z_{n+1} = \lambda z_n + z_n^2$ satisfy the equivalent properties (Bdd) and (Lin) above. Recall that a "random" number α is Diophantine of exponent $2 + \tau$ for $\tau > 0$ and hence satisfies (DC) almost surely.

On the other hand, it is easy to check that Cremer's condition (Cr) above is equivalent to

(Cr)' $$\sup_{k \geq 0} \frac{\log q_{k+1}}{2^{q_k}} = +\infty.$$

Under this condition, we know that the equivalent properties (Bdd) and (Lin) are not satisfied.

What about irrational numbers α which satisfy neither (DC) nor (Cr)? There is a rather large gap between the growth of the q_k implied by the two conditions: the first condition says $\log q_{k+1} < (1+\tau)\log q_k + C$ for some $C \in \mathbb{R}$ and all k, while the second implies that $\log q_{k+1} > 2^{q_k}$ infinitely often.

In 1965, Brjuno [2] proved that, if the convergents of α satisfy

(Br) $$\sum_{k \geq 0} \frac{\log q_{k+1}}{q_k} < +\infty,$$

then properties (Bdd) and (Lin) are still satisfied. Observe that this condition restricts the growth of the q_k much less than (DC); for instance, the condition $\log q_{k+1} = O(\sqrt{q_k})$ implies (Br) (since the q_k grow at least exponentially).

Brjuno's theorem is valid for the same kind of general recurrence relation $z_{n+1} = \lambda z_n + g(z_n)$ as Siegel's theorem (g complex differentiable in a neighbourhood of 0, $g(0) = 0$, $g'(0) = 0$).

On the other hand, in 1988 I proved the following result [10]:

Theorem 4. *Assume that Brjuno's condition (Br) is not satisfied:*

$$\sum_{k\geq 0} \frac{\log q_{k+1}}{q_k} = +\infty \,.$$

Then the properties (Bdd), (Lin) are not satisfied for the sequences defined by the quadratic relation $z_{n+1} = \lambda z_n + z_n^2$. *In particular, there exist initial values* z_0 *arbitrarily close to 0 such that the sequence* (z_n) *converges to* ∞.

For the quadratic recurrence relation $z_{n+1} = \lambda z_n + z_n^2$, we therefore know exactly for which irrational numbers α the equivalent properties (Bdd) and (Lin) are satisfied. Quadratic polynomials form the first family of dynamical systems in which explicit necessary and sufficient conditions are known for when a small divisor problem is stable, i.e., satisfies condition (Bdd).

However, this is not the end of the story. Replace the quadratic recurrence relation $z_{n+1} = \lambda z_n + z_n^2$ by any polynomial of degree $d \geq 3$ of the form

$$z_{n+1} = \lambda z_n + \sum_{2\leq \ell \leq d} f_\ell \, z_n^\ell$$

with $f_d \neq 0$. As before, write $\lambda = \exp(2\pi i \alpha)$. From Brjuno's theorem, we know that if α satisfies (Br), then properties (Bdd) and (Lin) are satisfied. On the other hand, it is conjectured that if α does not satisfy (Br), then properties (Bdd), (Lin) are not satisfied; but we do not have a proof at the moment.

More information on the topic of this section can be found in [7, Sec. 11].

5 Several Degrees of Freedom

Finally, we come back to the setting of planetary systems introduced in the first section, with one heavy central body (the sun) and $N-1$ planets orbiting around it. Consider a bounded solution of the unperturbed system (i.e., we do not take the mutual interaction between the planets into account). Each of the $N-1$ planets describes a Keplerian elliptic orbit with a period T_i ($1 \leq i \leq N-1$). Let $\omega_i = T_i^{-1}$, $1 \leq i \leq N-1$, be the corresponding frequencies. The *totally irrational* (or *non-resonant*) case occurs when there is no relation of the form

(Res) $$\sum_{i=1}^{N-1} k_i \, \omega_i = 0$$

with $k_i \in \mathbb{Z}$, not all 0.

One says that the frequency vector $\omega = (\omega_i)$ is *Diophantine* if there are constants $\gamma > 0$, $\tau \geq 0$ such that for any nonzero vector $k = (k_i) \in \mathbb{Z}^{N-1}$ one has

$$(\text{HDC})_{\gamma,\tau} \qquad \left| \sum_{i=1}^{N-1} k_i \, \omega_i \right| \geq \gamma \left(\sum_{i=1}^{N-1} |k_i| \right)^{2-N-\tau}.$$

We do not try to give a precise mathematical statement from KAM theory which applies in the setting of planetary systems. The general idea is that those solutions of the unperturbed system whose frequency vector is Diophantine (with $\tau > 0$ fixed, and γ not too small with relation to the size of the perturbation) will survive as slightly deformed quasiperiodic solutions of the perturbed system with the same frequency vector. Because a random frequency vector has a strictly positive probability to verify the required condition $(\text{HDC})_{\gamma,\tau}$ (the probability is strictly less than 1 because γ cannot be too small), it will also be true that a random initial condition for the differential equation of the planetary system leads with strictly positive probability to a quasiperiodic solution with a Diophantine frequency vector.

However, we expect that another set of initial conditions having strictly positive probability leads to solutions which are *not* quasiperiodic. To prove this statement and to understand these solutions is a major open problem.

References

[1] Jean-Benoît Bost, Tores invariants des systèmes dynamiques hamiltoniens (d'après Kolmogorov, Arnol'd, Moser, Rüssmann, Zehnder, Herman, Pöschel, ...) (French). Seminar Bourbaki 1984/85. *Astérisque* **133–134**, 113–157 (1986)
[2] Alexander D. Brjuno, Analytical form of differential equations. *Transactions of the Moscow Mathematical Society* **25**, 131–288 (1971); **26**, 199–239 (1972)
[3] Hubert Cremer, Über die Häufigkeit der Nichtzentren (German). *Mathematische Annalen* **115**, 573–580 (1938)
[4] Godfrey H. Hardy and Edward M. Wright, *An Introduction to the Theory of Numbers*, sixth edition. Oxford University Press, Oxford (2008)
[5] Alexander Khinchin, *Continued Fractions*. Translated from the third (1961) Russian edition. Dover, Mineola (1997)
[6] Serge Lang, *Introduction to Diophantine Approximations*, second edition. Springer, New York (1995)
[7] John Milnor, *Dynamics in One Complex Variable*, third edition. Princeton University Press, Princeton (2006)
[8] Henri Poincaré, *Les méthodes nouvelles de la mécanique céleste. Tome I. Solutions périodiques. Non-existence des intégrales uniformes. Solutions asymptotiques. Tome II. Méthodes de MM. Newcomb, Gyldén, Lindstedt et Bohlin. Tome III. Invariants intégraux. Solutions périodiques du deuxième genre. Solutions doublement asymptotiques* (French). First published in 1892–1899; Dover, New York (1957)
[9] Carl L. Siegel, Iteration of analytic functions. *Annals of Mathematics (2)* **43**, 607–612 (1942)
[10] Jean-Christophe Yoccoz, Théorème de Siegel, nombres de Bruno et polynômes quadratiques. Petits diviseurs en dimension 1 (French). *Astérisque* **231**, 3–88 (1995)

How do IMO Problems Compare with Research Problems?

Ramsey Theory as a Case Study

W. Timothy Gowers

Abstract. Although IMO contestants and research mathematicians are both attempting to solve difficult mathematical problems, there are important differences between their two activities. This is partly because most research problems involve university-level mathematical concepts that are excluded from IMO problems. However, there are more fundamental differences that are not to do with subject matter. To demonstrate this, we look at some results and questions in Ramsey theory, an area that has been a source both of IMO problems and of research problems.

1 Introduction

Many people have wondered to what extent success at the International Mathematical Olympiad is a good predictor of success as a research mathematician. This is a fascinating question: some stars of the IMO have gone on to extremely successful research careers, while others have eventually left mathematics (often going on to great success in other fields). Perhaps the best one can say is that the ability to do well in IMO competitions correlates well with the ability to do well in research, but not perfectly. This is not surprising, since the two activities have important similarities and important differences.

The main similarity is obvious: in both cases, one is trying to solve a mathematical problem. In this article, I would like to focus more on the differences, by looking at an area of mathematics, Ramsey theory, that has been a source both of olympiad problems and of important research problems.

W. Timothy Gowers
Department of Pure Mathematics and Mathematical Statistics, University of Cambridge, Cambridge CB3 0WB, UK. e-mail: W.T.Gowers@dpmms.cam.ac.uk

I hope to demonstrate that there is a fairly continuous path from one to the other, but that the two ends of this path look quite different.

Many expositions of Ramsey theory begin by mentioning the following problem.

Problem 1.1. *There are six people in a room, and any two of them are either good friends or bitter enemies. Prove that there must either be three people such that any two of them are good friends, or three people such that any two of them are bitter enemies.*

If you have not seen this problem (though I would imagine that most IMO contestants have encountered it), then you should solve it before reading on. It is not hard, but one learns a lot from working out the solution for oneself.

It is convenient to reformulate the problem before solving it, by stripping it of its irrelevant non-mathematical part (that is, its talk of people, friendship and enmity) and looking just at the abstract heart of the problem. One way of doing this is to represent the people by *points* in a diagram, and joining each pair of points by a (not necessarily straight) line. This gives us an object known as the *complete graph of order 6*. To represent friendship and enmity, we then colour these lines red if they join two people who are good friends and blue if they join two people who are bitter enemies. So now we have six points, with each pair of points joined by either a red line or a blue line. The standard terminology of graph theory is to call the points *vertices* and the lines *edges*. (These words are chosen because an important class of graphs is obtained by taking a polyhedron and forming a graph out of its vertices and edges. In such an example, there will be pairs of points not joined by edges, unless the polygon is a tetrahedron: these are therefore *incomplete* graphs.) Our task is now to prove that there must be a red triangle or a blue triangle, where a *triangle* in this context means a set of three edges that join three vertices.

To prove this, pick any vertex. It is joined by edges to five other vertices, so by the pigeonhole principle at least three of those edges have the same colour. Without loss of generality, this colour is red. There are therefore three vertices that are joined by red edges to the first vertex. If any two of these three vertices are joined by a red edge, then we have a red triangle. If not, then all three pairs of vertices from that group of three must be joined by blue edges, which gives us a blue triangle. QED

Let us define $R(k,l)$ to be the smallest number n such that if you colour each edge of the complete graph of order n red or blue, then you must be able to find k vertices such that any two of them are joined by a red edge, or l vertices such that any two of them are joined by a blue edge. We have just shown that $R(3,3) \leq 6$. (If you have not already done so, you should find a way of colouring the edges of the complete graph with five vertices red or blue in such a way that you do *not* get a red triangle or a blue triangle.)

It is not immediately obvious that the definition above makes sense: *Ramsey's theorem* is the assertion that $R(k,l)$ exists and is finite for every k and l.

A simple generalization of the argument we used to prove that $R(3,3) \leq 6$ can also be used to prove the following result, due to Erdős and Szekeres, which proves Ramsey's theorem and gives us information on how big $R(k,l)$ is.

Theorem 1.2. *For every k and l, we have the inequality*

$$R(k,l) \leq R(k-1,l) + R(k,l-1) .$$

Once again, if you have not seen this before, then you should prove it for yourself. (It is much easier than a typical IMO problem.) And you can then prove, by an easy inductive argument, that the inequality above implies that $R(k,l) \leq \binom{k+l-2}{k-1}$. (The main additional step is to note that $R(k,1) = 1$, or, if that bothers you, then the slightly safer $R(k,2) = k$, which allows you to get the induction started.)

This tells us that $R(3,4) \leq 10$. However, the true answer is in fact 9. Proving this is a more interesting problem — not too hard, but it involves an extra idea. From that and the inequality of Erdős and Szekeres, we may deduce that $R(4,4) \leq R(3,4) + R(4,3) = 18$, which turns out to be the correct answer: to show this you need to think of a red–blue colouring of the edges of the complete graph of order 17 such that no four vertices are all joined by red edges and no four vertices are all joined by blue edges. Such a graph exists, and it is rather beautiful: as ever, I would not want to spoil the fun by saying what it is.

We do not have to go much further than this before we enter the realms of the unknown. Using the Erdős–Szekeres inequality again we find that $R(3,5) \leq R(2,5) + R(3,4) = 5 + 9 = 14$, which turns out to be the actual value, and then that $R(4,5) \leq R(4,4) + R(3,5) = 32$. In 1995, McKay and Radziszowski showed, with a great deal of help from a computer, that in fact $R(4,5) = 25$. The best that is currently known about $R(5,5)$ is that it lies between 43 and 49.

It is not clear that the correct value of $R(5,5)$ will ever be known. Even if the answer is 43, a brute-force search on a computer through all of the $2^{\binom{43}{2}}$ red–blue colourings of the complete graph of order 43 would take far too long to be feasible. Obviously, there are ways of cutting this search down, but so far not by enough to make the computation feasible. At any rate, even if somebody does eventually manage to calculate $R(5,5)$, it is highly unlikely that $R(6,6)$ will ever be known. (It is known to be between 102 and 165.)

Why, you might ask, do we not try to find a *theoretical* argument rather than an ugly argument that checks huge numbers of graphs on a computer? The reason is that the largest graphs that avoid k vertices that are all joined by red edges or l vertices that are all joined by blue edges tend to be rather unstructured. In this respect, the graphs that demonstrate that $R(3,3) > 5$, $R(3,4) > 8$ and $R(4,4) > 17$ are rather misleading, since they have plenty of structure. This seems to be an example of the so-called "law of small numbers". (For a simpler example of this, take the fact that the first three primes, 2, 3 and 5, are consecutive Fibonacci numbers. This fact is of no significance whatsoever: there just aren't that many small numbers around so one expects coincidences.)

We therefore find ourselves in the unsatisfactory situation that there is probably no theoretical argument that gives an exact formula for $R(k,l)$, and therefore the best one can do is try to find clever search methods on a computer when k and l are small. This may sound a bit defeatist, but Gödel has taught us that we cannot just assume that everything we want to know has a proof. In the case of small Ramsey numbers, we do not learn anything directly from Gödel's theorem, since we could in principle calculate them by brute force, even if not in practice. However, the general message that nice facts do not have to have nice proofs still applies, and has an impact on the life of a research mathematician, which can be summed up in the following general problem-solving strategy, which I do not recommend to participants in mathematical olympiads.

Strategy 1.3. *When you are stuck on a problem, sometimes the best thing to do is give up.*

As a matter of fact, I do not entirely recommend it to research mathematicians either, unless it is coupled with the following rather more positive principle, which again I do not recommend to participants in mathematical olympiads.

Strategy 1.4. *If you cannot answer the question, then change it.*

2 Asymptotics of Ramsey Numbers

One of the commonest ways of changing a mathematical question when we find ourselves in a situation such as the one just described, faced with a quantity that we do not think we can calculate exactly, is to look for the best approximations that we can find, or at least to prove that the quantity must lie between L and U, where we try to make L (called a *lower bound*) and U (called an *upper bound*) as close as we can.

We have already obtained an upper bound for $R(k,l)$: the bound in question was $\binom{k+l-2}{k-1}$. For simplicity let us look at the case where $k = l$. Then we obtain the upper bound $\binom{2(k-1)}{k-1}$. Can we match that with a comparable lower bound?

Before we try to answer this question, we should first think about roughly how large $\binom{2(k-1)}{k-1}$ is. A fairly good approximation (but by no means the best known) is given by the formula $(k\pi)^{-1/2}4^{k-1}$, which we can think of as growing at about the same speed as 4^k (since the ratios of successive values of this function get closer and closer to 4).

This is a pretty large function of k. Is there any hope of finding a lower bound of anything like that size?

If by "finding" you mean writing down a rule that tells you when to colour an edge red and when to colour it blue, then the answer is that to find an exponentially large lower bound is a formidably difficult unsolved problem (though there are some fascinating results in this direction). However, in 1947, Erdős came up with a simple but revolutionary method of obtaining an exponentially large lower bound that does not involve finding one in this sense. Rather than give Erdős's proof, I shall just give the idea of the proof. It will be useful to introduce the following piece of terminology. If we have a red–blue colouring of the edges of the complete graph on n vertices, then let us call a set of vertices *monochromatic* if any two vertices in the set are joined by edges of the same colour.

Idea of Proof. *Do not attempt to find a colouring that works. Instead, choose the colours randomly and prove that the average number of monochromatic sets of size k is less than 1.*

If we can do that, then there must be a graph with no monochromatic sets of size k, since otherwise the average would have to be at least 1. The calculations needed to make this argument work turn out to be surprisingly simple, and they show that $R(k,k)$ is at least $\sqrt{2}^k$. (In fact, they give a slightly larger estimate than this, but not by enough to affect this discussion.)

The good news is that this lower bound is exponentially large. The bad news is that $\sqrt{2}^k$ is a *lot* smaller than 4^k. Can one improve one or other of these bounds? This is a central open problem in combinatorics.

Problem 2.1. *Does there exist a constant $\alpha > \sqrt{2}$ such that for all sufficiently large k we have the lower bound $R(k,k) \geq \alpha^k$, or a constant $\beta < 4$ such that for all sufficiently large k we have the upper bound $R(k,k) \leq \beta^k$?*

A more ambitious question is the following.

Problem 2.2. *Does the quantity $R(k,k)^{1/k}$ tend to a limit, and if so what is that limit?*

Probably $R(k,k)^{1/k}$ does tend to a limit. There are three natural candidates for what the limit might be: $\sqrt{2}$, 2 and 4. I have seen no truly convincing argument in favour of one of these over the other two.

There has been only a tiny amount of progress on these problems for several decades. So should we give up on them too? Definitely not. There is a profound difference between these extremely hard problems and the extremely hard problem of evaluating $R(6,6)$, which is that here one *expects* there to be a beautiful theoretical argument: it is just very hard to find. To give up the search merely because it is hard would be to go completely against the spirit of mathematical research. (Sometimes a single mathematician is well-advised to give up on a problem after spending a long time on it and getting nowhere. But here I am talking about a collective effort: pretty well all combinatorialists have at some time or another tried to improve the bounds for $R(k,k)$ and I am saying that this should continue until somebody eventually cracks it.)

3 What, in General, is Ramsey Theory?

A typical theorem in Ramsey theory concerns a structure that has many substructures that are similar to the main structure. It then says that if you colour the elements in the main structure with two colours (or more generally with r colours for some positive integer r), then you must be able to find a substructure all of whose elements have the same colour. For example, with Ramsey's theorem itself in the case where $k = l$, the structure is the complete graph of order $R(k,k)$ (or to be precise, the edges of the complete graph) and the substructures are all complete subgraphs of order k. Some Ramsey theorems also give information about how the size of the substructure depends on the size of the main structure and the number of colours.

Here is another example, a famous theorem of van der Waerden.

Theorem 3.1. *Let r and k be positive integers. Then there exists a positive integer n such that if you colour the numbers in any arithmetic progression X of length n with r colours, then you must be able to find some arithmetic progression Y inside X of length k such that all the numbers in Y have been given the same colour.*

I could talk a great deal about van der Waerden's theorem and its ramifications, but that would illustrate less well some of the more general points I want to make about IMO problems and research problems. Instead, I want to move in a different direction.

4 An Infinitary Structure and an Associated Ramsey Theorem

Up to now, the structures we have coloured — complete graphs and arithmetic progressions — have been finite. There is a version of Ramsey's theorem that holds for infinite complete graphs (another interesting exercise is to formulate this for yourself and prove it), but I want to look instead at a more complicated structure: the space of all infinite 01-sequences that are "eventually zero". An example of such a sequence is

00100111011000000000000000000000000000000000000000000000........

If s and t are two such sequences, and if the last 1 in s comes before the first 1 in t, then we write $s < t$. (One can think of this as saying that all the action in s has finished by the time the action in t starts.) If this is the case, then $s + t$ is another 01-sequence that is eventually zero. For example, you can add

00100111011000000000000000000000000000000000000000000000........

to

00000000000000011000110001100000000000000000000000000000........

and you will get the sequence

00100111011000110001100011000000000000000000000000000000........

Now let us suppose that we have sequences $s_1 < s_2 < s_3 < s_4 < \ldots$. That is, each s_i is a sequence of 0s and 1s, and all the 1s in s_{i+1} come after all the 1s in s_i. (Note that $(s_1, s_2, s_3, \ldots)$ is a sequence of sequences.) Therefore, if we take any sum of finitely many distinct sequences s_i, then we will obtain another sequence that belongs to our space. For example, we could take the sum $s_1 + s_2$, or the sum $s_3 + s_5 + s_6 + s_{201}$. The set of all possible sums of this kind is called the *subspace generated by* $s_1, s_2, s_3, \ldots$.

Now the entire space of sequences that we are talking about can be thought of as the subspace generated by the sequences $1000000\ldots$, $0100000\ldots$, $0010000\ldots$, $0001000\ldots$, and so on. Thus, the structure of the entire space is more or less identical to that of any of its subspaces. This makes it an ideal candidate for a Ramsey-type theorem. We can even guess what this theorem should say.

Theorem 4.1. *Let the 01-sequences that are eventually zero be coloured with two colours. Then there must be an infinite collection $s_1 < s_2 < s_3 < \ldots$ of sequences such that all the sequences in the subspace generated by the s_i have the same colour.*

That is, however you colour the sequences, you can find a sequence of sequences s_i such that s_1, s_2, $s_1 + s_2$, s_3, $s_1 + s_3$, $s_2 + s_3$, $s_1 + s_2 + s_3$, s_4 etc. all have the same colour.

This theorem is due to Hindman, and is too difficult to be thought of as an exercise. However, one thing that *is* a simple exercise is to show that once you have Hindman's theorem for two colours then you can deduce the same theorem for any (finite) number of colours.

Hindman's theorem is usually stated in the following equivalent form, which is easier to grasp, but which relates less well to what I want to talk about in a moment. Proving the equivalence is another exercise that is not too hard.

Theorem 4.2. *Let the positive integers be coloured with two colours. Then it is possible to find positive integers $n_1 < n_2 < n_3 < \ldots$ such that all sums of finitely many of the n_i have the same colour.*

This version of the theorem concerns addition. What if we try to introduce multiplication into the picture as well? Almost instantly we are back in the world of the unknown, since even the following innocent looking question is an unsolved problem.

Problem 4.3. *Let the positive integers be coloured with finitely many colours. Is it always possible to find integers n and m such that n, m, $n+m$ and nm all have the same colour? Is it even possible to ensure merely that $m+n$ and mn have the same colour (except in the trivial case $m = n = 2$)?*

This looks very much like an IMO problem. The difference is that it just happens to be far far harder (and one does not have the helpful knowledge that somebody has solved it and deemed it suitable for a mathematics competition).

5 From Combinatorics to Infinite-dimensional Geometry

We represent three-dimensional space by means of coordinates. Once we have done this, it is easy to define d-dimensional space for any positive integer d. All we have to do is express our concepts in terms of coordinates and then increase the number of coordinates. For example, a four-dimensional cube could be defined as the set of points (x, y, z, w) such that each of x, y, z and w is between 0 and 1.

If we want to (which we often do when we are doing university-level mathematics), we can even extend our concepts to *infinite*-dimensional space. For instance, an infinite-dimensional sphere of radius 1 can be defined as the set of all sequences $(a_1, a_2, a_3, \ldots)$ of real numbers that satisfy the condition $a_1^2 + a_2^2 + a_3^2 + \cdots = 1$. (Here I am using the word "sphere" to mean the surface of a ball rather than a solid ball.)

In our infinite-dimensional world, we also like to talk about lines, planes, and higher-dimensional "hyperplanes". In particular, we are interested in infinite-dimensional hyperplanes. How do we define these? Well, a plane going through the origin in three-dimensional space can be defined by taking two points $\mathbf{x} = (x_1, x_2, x_3)$ and $\mathbf{y} = (y_1, y_2, y_3)$ and forming all combinations $\lambda \mathbf{x} + \mu \mathbf{y}$ of these two points. (Here, $\lambda \mathbf{x} + \mu \mathbf{y}$, when written out in coordinates, is $(\lambda x_1 + \mu y_1, \lambda x_2 + \mu y_2, \lambda x_3 + \mu y_3)$.) We can do something similar in infinite-dimensional space. We take a sequence of points $\mathbf{p}_1, \mathbf{p}_2, \mathbf{p}_3, \ldots$ (here each $\mathbf{p}_i$ will itself be an infinite sequence of real numbers) and we take all combinations (subject to certain technical conditions) of the form $\lambda_1 \mathbf{p}_1 + \lambda_2 \mathbf{p}_2 + \lambda_3 \mathbf{p}_3 + \cdots$.

It turns out that if we look at the intersection of an infinite-dimensional sphere with an infinite-dimensional hyperplane, then we get another infinite-dimensional sphere. (Apart from the fact that all the dimensions are infinite, this is a bit like the fact that if you intersect a sphere with a plane then you get a circle.) Let us call this a *subsphere* of the original sphere. Once again we seem to be ideally placed for a Ramsey-type theorem, since we have a structure (a sphere) with many substructures (subspheres) that look exactly like the structure itself. Suppose that we colour an infinite-dimensional sphere with two colours. Can we always find a subsphere that has been coloured with only one colour?

There is some reason to expect a result like this to be true. After all, it is quite similar to Hindman's theorem, in that both statements involve colouring some infinite-dimensional object, defined by coordinates, and looking for a monochromatic infinite-dimensional subobject of a similar type. It is just that in Hindman's theorem all the coordinates have to be 0 or 1.

Unfortunately, however, the answer to our new question is no. If **p** belongs to a subsphere, then −**p** must belong to the same subsphere. So we could colour **p** red if its first non-zero coordinate is positive and blue if its first non-zero coordinate is negative. In that case, **p** and −**p** will always receive different colours. (Since the squares of their coordinates have to add up to 1 they cannot all be zero.)

This annoying observation highlights another difference between IMO problems and the kinds of questions that come up in mathematical research.

Principle 5.1. *A significant proportion of conjectures that come up naturally in one's research turn out to be easy or badly formulated. One has to be lucky to stumble on an interesting problem.*

However, under these circumstances we can apply a variant of a strategy I mentioned earlier.

Strategy 5.2. *If the question you are thinking about turns out to be uninteresting, then change it.*

Here is a small modification to the problem about colouring spheres, which turns it from a bad problem into a wonderful one. Let us call a subsphere c-*monochromatic* if there is a colour such that every point in the subsphere is within a distance c from some point of that colour. We think of c as small, so what this is basically saying is that we do not ask for all points in the subsphere to be red (say), but merely that every point in the subsphere is *close* to a red point.

Problem 5.3. *If the infinite-dimensional sphere is coloured with two colours, and c is a positive real number, then is it always possible to find a c-monochromatic infinite-dimensional subsphere?*

This problem remained open for a long time and became a central question in the theory of Banach spaces, which are a formalization of the idea of infinite-dimensional space and one of the central concepts in research-level mathematics. Unfortunately, it too had a negative answer, but the counterexample that shows it is *much* more interesting and *much* less obvious than the counterexample to the bad version of the problem. It was discovered by Odell and Schlumprecht.

The example of Odell and Schlumprecht killed off the hope of a Hindman-like theorem for Banach spaces (except for one particular space where the similarity to the space of 01-sequences is more pronounced, for which I obtained such a theorem). However, it did not entirely destroy the connections between Ramsey theory and Banach-space theory, as we shall see in the next section.

Before we finish this section, let me mention another difference between IMO problems and research problems.

Principle 5.4. *A research problem can change from being completely out of reach to being a realistic target.*

To somebody with experience only of IMO problems, this may seem strange: how can the difficulty of a problem change over time? But if you look back at your own mathematical experience, you will know of many examples of problems that "became easy". For example, consider the problem of finding the positive real number x such that $x^{1/x}$ is the biggest it can be. If you know the right tools, then you argue as follows. The logarithm of $x^{1/x}$ is $\log x/x$, and the logarithm function is increasing, so the problem is equivalent to maximizing $\log x/x$. Differentiating gives us $(1 - \log x)/x^2$, which is zero only when $x = e$, and decreasing there. Hence, the maximum is at $x = e$.

That solution is fairly straightforward, both to understand and to find in the first place, but only if one knows a bit of calculus. So the problem is out of reach to people who do not know calculus and a realistic target to those who do. Something similar to this happens in mathematical research, but the additional point I am making is that it can be a *collective* phenomenon and not just an individual one. That is, there are many problems that are out of reach simply because the right technique has not been invented yet.

You might object that that does not really mean that the problem is out of reach: it just means that part of the work of solving it is to invent the right technique. In a way that is true, but it overlooks the fact that mathematical techniques are very often used to solve problems that were not the problems that originally motivated the technique. (For instance, Newton and Leibniz did not invent calculus so that we could maximize the function $x^{1/x}$.) Thus, it may well happen that Problem B becomes a realistic target because somebody has invented the right technique while thinking about Problem A.

I mention all this here because Odell and Schlumprecht built their counterexample by modifying (in a very clever way) an example that Schlumprecht had built a few years earlier for a completely different reason.

6 A Little Bit More About Banach Spaces

I am aware that I have not explained very clearly what a Banach space is, and I may have given the impression that the only notion of distance in infinite-dimensional spaces is what you get by generalizing Pythagoras's theorem and defining the distance of a point $(a_1, a_2, a_3, \ldots)$ from the origin to be $\sqrt{a_1^2 + a_2^2 + a_3^2 + \ldots}$.

However, other notions of distance are possible and useful. For instance, for any $p \geq 1$ we can define the distance from $(a_1, a_2, a_3, \ldots)$ to the origin to be the p-th root of $|a_1|^p + |a_2|^p + |a_3|^p + \ldots$. Of course, there will be sequences for which this number is infinite. We regard these sequences as not belonging to the space.

It is not obvious in advance that this will be a sensible notion of distance, but it turns out to have some very good properties. Writing **a** and **b** for the sequences $(a_1, a_2, a_3, \ldots)$ and $(b_1, b_2, b_3, \ldots)$, and writing $\|\mathbf{a}\|$ and $\|\mathbf{b}\|$ for the distances from **a** and **b** to the origin, usually known as the *norms* of **a** and **b**, we can express these properties as follows.

(i) $\|\mathbf{a}\| = 0$ if and only if $\mathbf{a} = (0, 0, 0, \ldots)$.
(ii) $\|\lambda \mathbf{a}\| = |\lambda| \cdot \|\mathbf{a}\|$ for every **a**.
(iii) $\|\mathbf{a} + \mathbf{b}\| \leq \|\mathbf{a}\| + \|\mathbf{b}\|$ for every **a** and **b**.

All three of these properties are properties that we are familiar with from the usual notion of distance in space. (Note that we can define the distance from **a** to **b** to be $\|\mathbf{a} - \mathbf{b}\|$.) A *Banach sequence space* is a set of sequences together with some norm that satisfies properties (i)–(iii) above, together with a more technical condition (called *completeness*) that I shall not discuss.

The particular example where $\|\mathbf{a}\|$ is defined to be $\left(\sum_{n=1}^{\infty} a_n^2\right)^{1/2}$ is a very special kind of Banach space called a *Hilbert space*. I will not say what a Hilbert space is, except to say that it has particularly good symmetry properties. One of these good properties is that every subspace of a Hilbert space is basically just like the space itself. We have already seen this: when we intersected an infinite-dimensional sphere with an infinite-dimensional hyperplane we obtained another infinite-dimensional sphere. This property, that all subspaces are "isomorphic" to the whole space, does not seem to hold for any other space, so Banach himself asked the following question in the 1930s.

Problem 6.1. *Is every space that is isomorphic to all its (infinite-dimensional) subspaces isomorphic to a Hilbert space?*

To put that more loosely, is a Hilbert space the only space with this particularly good property? The difficulty of the question is that there are many ways that two infinite-dimensional spaces can be isomorphic, so ruling them all out for a non-Hilbert space and some carefully chosen subspace is likely to be hard.

This is another example of a problem that turned from impossible to possible as a result of developments connected with other problems, and I was lucky enough to be in the right mathematical place at the right time, so to speak. Some work of Komorowski and Tomczak–Jaegermann (incidentally, I am mentioning several mathematicians whose names will mean very little to most readers of this article, but decided against prefacing every single one with "a mathematician called") showed that if there was a counterexample to the problem, it would have to be rather nasty in a certain sense.

Now it is far from obvious that there could be a space as nasty as what would be required, but it so happened that a couple of years earlier Maurey and I had constructed just such a nasty space, and our nasty space was *so* nasty that for entirely different reasons it had no chance of being a counterexample to Banach's question. This raised the possibility that the answer to Banach's question was yes, because nice examples couldn't work, and nasty examples couldn't work either. In order to make an approach like this work, I found myself needing to prove a statement of the following kind.

Statement 6.2. *Every infinite-dimensional Banach space has an infinite-dimensional subspace such that all its subspaces are nice or all its subspaces are nasty.*

Now this has strong overtones of Ramsey theory: we could think of nice subspaces as "red" and nasty subspaces as "blue".

7 A Weak Ramsey-type Theorem for Subspaces

There is, however, one important difference between Statement 6.2 and our earlier Ramsey-theoretic statements, which is that here the objects we are colouring are (infinite-dimensional) *subspaces* rather than *points*. (However, I should point out that in Ramsey's theorem itself we coloured edges rather than vertices, so the idea of colouring something other than points is not completely new.) How do we fit this into our general framework?

It is in fact not too hard. The structures we are colouring can be thought of as "the structure of all subspaces of a given space". If we take any subspace, then all *its* subspaces form a structure of a similar kind to the structure we started with, so we can think of trying to prove a Ramsey-type theorem.

The best we could hope to prove would be something like this: if you colour all the subspaces of some space red or blue, then there must be a subspace such that all of *its* subspaces have the same colour. However, not too surprisingly, this turns out to be far too much to hope for, for both boring and interesting reasons. The boring reason is similar to the reason that we could not colour the points of an infinite-dimensional sphere and hope for a monochromatic infinite-dimensional subsphere. The interesting reason is that even if we modify the statement so that we are looking for subspaces that are *close* to all being the same colour (in some suitable sense of "close"), the results of Odell and Schlumprecht, which concerned colouring points, can be used fairly easily to show that we will not necessarily find them.

We appear to have reached a dead end, but in fact we have not, because for the application I had in mind, I did not need the full strength of a Ramsey theorem. Instead, I was able to get away with a "weak Ramsey theorem", which I shall now briefly describe.

To do so, I need to introduce a curious-looking game. Suppose that we are given a collection Σ of sequences of the form $(\mathbf{a}_1, \mathbf{a}_2, \mathbf{a}_3, \ldots)$, where all $\mathbf{a}_i$ are points in a Banach space. (It is important, here and in many previous places in this article, to keep in mind what the objects are that I am talking about. This can get quite complicated: Σ is a collection of sequences, as I have just said; but the terms in each sequence are themselves points in a Banach space, so they are sequences of real numbers, which is why I have written them in bold face. Thus, Σ is a set of sequences of sequences of real numbers. One could take this even further and say that each real number is represented by an infinite decimal, so Σ is a set of sequences of sequences of sequences of numbers between 0 and 9. But it is probably easier to think of the terms $\mathbf{a}_n$ as points in an infinite-dimensional space and forget about the fact that they have coordinates.) Given the collection Σ, Players A and B then play as follows. Player A chooses a subspace S_1. Player B then chooses a point $\mathbf{a}_1$ from S_1. Player A now chooses a subspace S_2 (which does not have to be a subspace of S_1) and player B chooses a point $\mathbf{a}_2$ from S_2. And so on. At the end of this infinite process, player B will have chosen a sequence $(\mathbf{a}_1, \mathbf{a}_2, \mathbf{a}_3, \ldots)$. If this sequence is one of the sequences in the collection Σ, then B wins, and otherwise A wins.

Now obviously who wins this game depends heavily on what Σ is. For example, if there happens to be a subspace S such that it is impossible to find points $\mathbf{a}_n$ in S that form a sequence $(\mathbf{a}_1, \mathbf{a}_2, \mathbf{a}_3, \ldots)$ in Σ, then A has the easy winning strategy of choosing S every single time, but if Σ contains almost all sequences then B will be expected to have a winning strategy.

Here, then, is the weak Ramsey theorem that turned out to be enough to prove a suitably precise version of Statement 6.2 and hence answer Banach's question (Problem 6.1). I have slightly oversimplified the statement. Before I give the statement itself, let us make the following definition. If S is a subspace, then the *restriction of the game to S* is the game that results if all the subspaces $S_1, S_2, \ldots$ chosen by A have to be subspaces of S (and hence all the points chosen by B have to be points in S).

Theorem 7.1. *For every collection Σ of sequences in a Banach space there is a subspace S such that either B has a winning strategy for the restriction of the game to S or no sequence in Σ can be made out of points of S.*

To see how one might call this a weak Ramsey theorem, let us colour a sequence red if it belongs to Σ and blue otherwise. Then the theorem says that we can find a subspace S such that either all the sequences built out of points in S are blue, or there are so many red sequences built out of points in S that if the game is confined to S then B has a winning strategy for producing red sequences.

In other words, we have replaced "all sequences in S are red" by "so many sequences in S are red that B has a winning strategy for producing them".

It is one thing to formulate such a statement and observe that it is sufficient for one's purposes, but quite another to prove it. This brings me to another difference between IMO problems and research problems, which is that the following problem-solving strategy is far more central to research problems than to IMO problems.

Strategy 7.2. *If you are trying to prove a mathematical statement, then search for a similar statement that has already been proved, and try to modify the proof appropriately.*

I would not want to say that this always works in research or that it never works in an IMO problem, but with IMO problems it is much more common to have to start from scratch.

Going back to the weak Ramsey theorem, it turned out to resemble another infinitary Ramsey theorem, due to Galvin and Prikry. The resemblance was close enough that I was able to modify the argument and prove what I needed. And luckily I had been to a course in Cambridge a few years earlier in which Béla Bollobás had covered the theorem of Galvin and Prikry.

8 Conclusion

I do not have much to say in conclusion that I have not already said. However, there is one further point that is worth making. If you are an IMO participant reading this, it may seem to you that your talent at solving olympiad problems has developed almost without your having to do anything: some people are just good at mathematics. But if you have any ambition to be a research mathematician, then sooner or later you will need to take account of the following two principles.

Principle 8.1. *If you can solve a mathematical research problem in a few hours, then it probably wasn't a very interesting problem.*

Principle 8.2. *Success in mathematical research depends heavily on hard work.*

Even from the examples I have just given, it is clear why. When one sets out to solve a genuinely interesting research problem, one usually has only a rather hazy idea of where to start. To get from that hazy idea to a clear plan of attack takes time, especially given that most clear plans of attack have to be abandoned anyway — for the simple reason that they do not work. But you also need to be ready to spot the connections and similarities to other problems, and to have developed your own personal toolbox of techniques, bits of mathematical knowledge, and so on. Behind any successful research mathematician will be thousands of hours spent pondering mathematics, only very few of which will have directly led to breakthroughs. It is strange, in a way, that anybody is prepared to put in those hours. Perhaps it is because of a further principle such as this.

Principle 8.3. *If you are truly interested in mathematics, then hard mathematical work does not feel like a chore: it is what you want to do.*

Further Reading

[1] Ron Graham, Bruce Rothschild, and Joel Spencer, *Ramsey Theory*. Wiley-Interscience Series in Discrete Mathematics and Optimization. Wiley, New York (1990)
This book contains a wealth of material about Ramsey's theorem, van der Waerden's theorem, Hindman's theorem, and many other results. It is a highly recommended starting point for anyone interested in the subject.

[2] Béla Bollobás, *Linear Analysis: An Introductory Course, second edition*. Cambridge University Press, Cambridge (1999), xii+240 pp.
Banach spaces belong to a branch of mathematics known as Linear Analysis. This is an introduction to that area that is likely to appeal to IMO contestants. (Look out for the exercise that has two stars ...)

[3] Edward Odell and Thomas Schlumprecht, The distortion problem. *Acta Mathematica* **173**, 259–281 (1994)
This publication contains the example of Odell and Schlumprecht that is mentioned in Section 5.

[4] W. Timothy Gowers, An infinite Ramsey theorem and some Banach-space dichotomies. *Annals of Mathematics (2)* **156**, 797–833 (2002)
This publication contains my result about the infinite game and its consequences.

The above two publications assume familiarity with Banach spaces and so will be very tough going for a pre-university reader. Another possibility that might be slightly better is the following survey paper I wrote about the connections between Ramsey theory and Banach spaces.

[5] W. Timothy Gowers, Ramsey methods in Banach spaces. In: William B. Johnson and Joram Lindenstrauss (editors), *Handbook of the Geometry of Banach Spaces, volume 2*, pp. 1071–1097. North-Holland, Amsterdam (2003)

If you can get hold of the first volume as well, then the first chapter is about basic concepts of the subject, which might also be helpful; it is the following:

[6] William B. Johnson and Joram Lindenstrauss, Basic concepts in the geometry of Banach spaces. In: William B. Johnson and Joram Lindenstrauss (editors), *Handbook of the Geometry of Banach Spaces, volume 1*, pp. 1–84. North-Holland, Amsterdam (2001)

How do Research Problems Compare with IMO Problems?

A Walk Around Games

Stanislav Smirnov

Abstract. Are the problems one encounters at IMOs and the problems one encounters as a research mathematician alike? We will make use of a few examples to show their similarities as well as their differences. The problems chosen come from different areas, but are all related to arrangements of numbers or colors on graphs, and to games one can play with them.

1 Do Mathematicians Solve Problems?

When asked what research in mathematics is like, mathematicians often answer: *We prove theorems.* This best describes the quintessential part of mathematical work and also how it differs from research in, say, biology or linguistics. And though in school one often gets the impression that all theorems were proved ages ago by Euclid and Pythagoras, there are still many important unsolved problems.

Indeed, research mathematicians do solve problems. There are other important parts of research, from learning new subjects and looking for connections between different areas to introducing new structures and concepts and asking new questions. Some even say that posing a problem is more important than solving it. In any case, without problems there would be no mathematics, and solving them is an important part of our job. As Paul Halmos, who has written several books about research problems, once said: *Problems are the heart of mathematics.*

Students often ask: *How does doing research compare to the IMO experience?* There are many similarities, and problem solving skills certainly help in research, so many IMO competitors go on to become mathematicians. There

---

Stanislav Smirnov
Section de Mathématiques, Université de Genève, 2–4 rue du Lièvre, CP 64, 1211 Genève 4, Switzerland. e-mail: Stanislav.Smirnov@unige.ch

are also, however, some differences. So how do IMO problems compare to research problems?

Substantial differences in solutions are often mentioned. Typically, an IMO problem will have a nice solution that requires the use of a limited number of methods (and hopefully at least one participant can find it within the given four and a half hours). Problems one encounters as a mathematician often require methods from very different mathematical areas, so that ingenuity alone would not suffice to solve them. Moreover for many problems that are easy to formulate only long and technical solutions have been found; it may even be the case that no nice solution exists. And when starting to work on a research problem, you cannot be sure that there is a solution at all. So you do not need to be as quick as when at IMO competitions, but you need to have much more determination — you rarely prove a theorem in four hours, and sometimes it takes years to advance on an important question. On the positive side, mathematics is now more of a collective effort, and collaborating with others is a very rewarding experience.

Not only the solutions, but also the problems themselves are somewhat different. Three IMO problems easily fit on one sheet of paper, but it takes much more to describe most of the open questions in mathematical research. Fortunately, there are exceptions that have equally short formulations, and mathematicians very much enjoy tackling these — they often play a catalytic role, attracting our attention to a particular area. Motivation for posing problems also differs. Whereas many research problems (like most IMO problems) are motivated by the inner beauty of mathematics, a significant number originate in physical or practical applications, and then the questions asked change somewhat.

So, are research problems and IMO problems really different? I would say that they have more in common, and that mathematicians enjoy beautiful problems, elegant solutions, and the process of working on a problem just as much as IMO contestants do.

To highlight both the similarities and the differences, I describe below a few problems that I have encountered and that would do well as both IMO and research questions (but would come in a slightly different light). While coming from different areas of mathematics, all of these are concerned with numbers (or colors) placed on a graph.

2 The Pentagon Game

The Pentagon Game is one of the most memorable problems I solved at olympiads. It was proposed by Elias Wegert of Germany, who also took part in the 50-th IMO as a coordinator:

> **27th International Mathematical Olympiad**
> Warsaw, Poland
> Day I
> July 9, 1986
>
> **Problem 3.** *To each vertex of a regular pentagon an integer is assigned in such a way that the sum of all five numbers is positive. If three consecutive vertices are assigned the numbers x, y, z respectively and $y < 0$ then the following operation is allowed: the numbers x, y, z are replaced by $x+y$, $-y$, $z+y$ respectively. Such an operation is performed repeatedly as long as at least one of the five numbers is negative. Determine whether this procedure necessarily comes to an end after a finite number of steps.*

I was among the students, and it was a very nice problem to tackle, perhaps the hardest at that IMO. It is almost immediately clear that one should find some positive integer function of a configuration that decreases with each operation. Indeed, two such semi-invariants were found by participants, and since we cannot decrease a positive integer infinitely many times, the procedure will necessarily come to an end.

This is a classical *combinatorics* problem, and if you are into olympiads, you certainly have seen a few very similar ones. What is interesting is that its life was more like that of a research problem. It was originally motivated by a question that arose in research dealing with partial reflections of polygons. So even the motivating area, *geometry*, was very different.

The combinatorial structure of this game is interesting in itself, and studying it on graphs that are different from a pentagon could have led to a few IMO problems and perhaps a research paper. But connections with algebraic questions have surfaced, which made it much more interesting for mathematical research. I was very pleasantly surprised to hear a talk that originated from the Pentagon Game at a research seminar some twenty years after that IMO. The talk was by Qëndrim Gashi, who used a version of the game due to Shahar Mozes to prove the Kottwitz-Rapoport conjecture in *algebra*. So far, versions of the Pentagon Game have led to more than a dozen research papers — not bad for an IMO problem!

These kinds of unexpected links between different areas, and between simple and complicated subjects, are one of the best things about doing mathematical research. Unfortunately, they often pass unnoticed in IMO competitions.

3 The Game of Life

There are many similar games with numbers, and they may yield much wider connections, often stretching beyond mathematics.

The most famous is perhaps John Conway's *Game of Life*. This is an example of a very rich class of games called cellular automata, first introduced by John von Neumann and Stanisław Ulam. In such games the graph is taken to be a regular grid, only a finite set of numbers (or states) is used, and an operation consists of *simultaneously* changing all numbers according to some rule depending on their neighbors.

The Game of Life is played on a square grid, with cells (squares) that can have two states: 1 and 0. The operation simultaneously changes the states of all cells by a simple rule depending on the state of their eight neighbors (i.e. squares with which they share an edge or a vertex). The rule is usually formulated in terms of live (state 1) and dead (state 0) cells:

- a live cell with 2 or 3 live neighbors stays alive,
- a live cell with < 2 live neighbors dies as if from loneliness,
- a live cell with > 3 live neighbors dies as if from overcrowding,
- a dead cell with 3 live neighbors becomes alive,
- a dead cell with $\neq 3$ live neighbors stays dead.

The rule is very simple, but it leads to rather complicated phenomena. Besides configurations that stay fixed (e.g. a 2×2 square of live cells) and those that oscillate periodically (e.g. a 1×3 rectangle of live cells), there are configurations that exhibit nontrivial behavior. For example, the "glider" pattern moves one step southeast every four operations, whereas "Bill Gosper's gun" shoots out a new glider every thirty operations. Patterns like this allow us to use the Game of Life to even model a computer, though the needed configurations would be rather large and complicated. Also, chaotic configurations quite often transform into complex patterns with some structure, which makes the game interesting to scientists in other disciplines, from philosophy to economics.

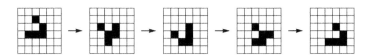

Fig. 1. A glider configuration: every four operations, it moves one step southeast. Live cells are colored black, while dead cells are colored white.

The Game of Life was popularized by Martin Gardner well beyond the mathematics community, and one can now easily find information about it, including interactive models, on the internet (it is quite instructive

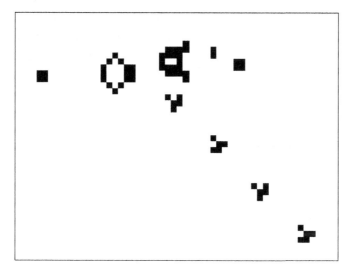

Fig. 2. Gosper's glider gun configuration: it shoots out a glider every thirty operations. There are also configurations that reflect, turn, or destroy gliders. Together, they can be used to build very complicated structures, and even model a computer.

and amusing to watch the Game of Life in "real time", for instance on http://www.bitstorm.org/gameoflife). Moreover many questions about this game could well double as IMO and research problems, and there are many other interesting cellular automata.

4 The Sandpile Model

It seems natural that if one wants to model more phenomena, some randomness has to be added, so that the evolution is not uniquely determined by the initial configuration. Indeed, it has been known for a long time that by introducing randomness into simple games (roughly speaking, we take two or more rules, and at each vertex toss a coin to decide which one to apply) one can accurately model many phenomena exhibiting phase transitions — from ferromagnetic materials to the spread of epidemics. It came as quite a surprise, however, that similar phenomena could be observed even in usual, non-random games.

One famous such game, the *Sandpile Model*, was introduced by three physicists, Per Bak, Chao Tang, and Kurt Wiesenfeld, in 1987. The game is played on an infinite square grid by writing positive integers on finitely many cells, and zeroes elsewhere. These integers are thought of as heights of a pile of sand. (One can also play on a finite region, but then one cell is designated a "pit": all grains falling there disappear.)

In the original model, all cells were changed simultaneously. Below we give a modified version, due to Deepak Dhar, where the same rule is applied but only to one cell at a time, much like in the Pentagon Game. The operation is slightly different, though: whereas in the Pentagon Game we subtract $2y$ from a vertex with value y and redistribute this amount evenly among neighbors, here we subtract 4. To be precise, the operation in the Sandpile Model is the following: if some cell with h grains is too tall (i.e. $h \geq 4$), it will topple over, giving one grain of sand to each of its four neighboring cells (i.e. those cells with which it shares an edge), which have, say, h_1, h_2, h_3, and h_4 grains. Thus the operation is described by

$$h \to h - 4,$$
$$h_j \to h_j + 1.$$

Like in the Pentagon Game, the operation is performed repeatedly as long as we can find a cell with $h \geq 4$. Eventually, we stop, reaching a stable configuration with all piles satisfying $h \leq 3$. The sequence of operations that leads to a stable configuration is called an *avalanche*.

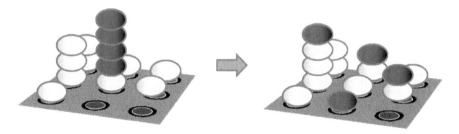

Fig. 3. A pile of five grains is toppled, with one grain going to each of its four neighbors. Note that we created a new pile of four grains, ready to be toppled. (The figure shows a 3×3 square within the infinite square grid.)

In order to work with the sandpile model, one first has to solve a problem very similar to one from the 1986 IMO:

Show that an avalanche ends after a finite number of operations.

Often, more than one pile has big height, so that we must choose the pile that we wish to topple. But it appears (unlike with the Pentagon Game) that

at the end of an avalanche, one obtains the same configuration independently of the order of operations.

Can you prove these two statements? In addition to being important lemmas in research papers, they would make for nice IMO problems.

Fig. 4. 50 000 grains were added at the central square, resulting in an avalanche. This is a pile obtained afterwards, with colors (white, yellow, orange, red) representing the heights (0, 1, 2 or 3) of the cells. The shape is almost circular. If we keep adding more grains, will it look more like a circle?

According to physicists (and we have high respect for our colleagues — interactions between mathematics and physics have enriched both fields), the really interesting problems only start here. Once the avalanche comes to an end, we can add one more grain of sand at some fixed center cell (or at a random place). This leads to a new avalanche. Then we add a new grain, and so forth.

When the original sandpile paper appeared, physicists were struggling to explain two recurring phenomena in nature: the "$1/f$ noise" and the appearance of spatial fractal structures. Both phenomena are often encountered in everyday life: $1/f$-noise (so called because its power is inversely proportional to its frequency) appears in areas as different as hissing sounds of a home stereo system, human heartbeats, or stock market fluctuations. And seemingly chaotic yet self-similar fractal structures (so called because they behave as if their dimension was fractional) can be seen in the shapes of clouds, systems of blood vessels, or mountain ranges. Based on physical observations,

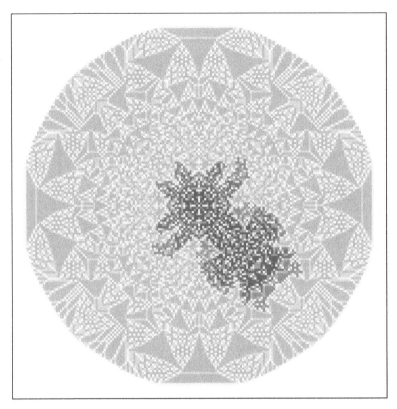

Fig. 5. An avalanche triggered by adding one grain (not at the center) to the pile of 50 000 grains from Figure 4. Highlighted are cells where many topplings occured; other cells are deemphasized. What would be the average size of such an avalanche?

one could ask questions like: given a pile totaling N grains, what would be the average diameter (size), length (number of grains toppled), or shape of an avalanche?

Computer experiments have exposed both phenomena in the Sandpile Model: adding grains to a stable configuration triggers an avalanche of a fractal shape and a size distributed not unlike $1/f$ noise. Moreover it is usually the case that adding one grain of sand either does not change much or causes almost the entire pile to collapse in an avalanche. Such behavior is characteristic of physical systems at "critical points", like a liquid around freezing temperature, when a small change (slightly decreasing the temperature, or sometimes dropping a small crystal into it) can cause it to freeze. However, the Sandpile Model is attracted to the critical point, whereas most physical systems are difficult to keep at criticality. And the sandpile model, though very simple to formulate, turned out to be the first mathematical example of what physicists now call "self-organized criticality".

Despite extensive computer simulations providing convincing evidence, as well as a vast literature, most of these questions remain open after 20 years, but mathematicians work for much longer than four and a half hours! Also, it is not clear that such questions will have a nice (and provable) answer, and their original motivation comes from outside of mathematics, so that they would likely not be asked as IMO problems.

Though the original motivation for the Sandpile Model came from physics, mathematicians have since asked a number of mathematical questions about it, motivated by the simplicity and beauty of the model. Some of these questions are of geometric nature and would do fine as IMO problems if only we had nice solutions. For example, we can keep adding particles at the origin, and the pile will grow in size. Will it look like a circle, as Figure 4 suggests? Apparently not — it seems that it will start developing sides after a while. So what will be its shape? How can we describe the intricate patterns we see? Despite much work, we still do not know.

5 The Self-Avoiding Walk

This article started with a problem that I solved at the 1986 IMO. It therefore seems appropriate to end with one of the problems that I am trying to solve now. Lately, I have spent a lot of time working on a large class of questions about systems undergoing phase transitions, and they too can be formulated as games played on a grid. The most famous one is perhaps the Ising Model, which is applied to many phenomena, from the magnetization of metals to the neuron activity in the brain. It can be formulated similarly to the Game of Life: again the cells of the square lattice can have two states (+ and − polarization for magnets, active and passive for neurons). The operation also counts the number of + neighbors, but then one makes a coin toss to decide the outcome. The coin is biased, so that the cell is more likely to choose the same state as the majority of its neighbors.

Interestingly, when the bias is gradually changed, the generic state of the Ising model undergoes a phase transition from the chaotic state to the ordered state (when the majority of the cells tends to be in the same state). There is also a deterministic version which only uses randomness to generate the initial state.

There is a number of similar problems, and below I describe one which is perhaps the simplest to formulate. One does not even need to define a game — it is sufficient to count configurations on a grid. Moreover, we find here a fruitful interaction between chemistry, physics, and mathematics!

In the 1940s, Paul Flory, a Nobel Prize winning chemist, asked how a polymer is positioned in space. He proposed modeling polymer chains by broken lines drawn on a grid without self-intersections (since a molecule obviously won't intersect itself). Equivalently, imagine a person walking on a grid in such a way that he does not visit the same vertex twice. We call this a *self-avoiding walk*. Each n-step trajectory would then model a possible position of a length n chain.

The fundamental question is what generic chains would look like, but before answering this one has to ask the following question:

> *How many length n self-avoiding walks starting from the origin can one draw on a grid?*

Denote this number by $C(n)$; walks that are rotations of each other are counted separately. It depends on the grid chosen, and in general we do not expect to have a nice formula for it (though it may exist despite expectations — sometimes miracles do happen). One thus asks how fast this number grows in terms of n. An IMO-type problem might be the following:

> *Show that there is a constant μ such that the number of self-avoiding walks satisfies $C(n) \approx \mu^n$ as n tends to ∞.*

The sign $\approx$ above means that however small we take ε, for large enough n we have $(\mu - \varepsilon)^n < C(n) < (\mu + \varepsilon)^n$. The problem above is not difficult and follows from the observation that a self-avoiding walk of length $n + m$ can be cut into two self-avoiding walks of lengths n and m. Indeed, its first n steps give a self-avoiding walk (starting at the origin), while its last m steps give a self-avoiding walk starting at the end of the n-th step (this walk can be translated so that it starts at the origin). Hence $C(n+m) \leq C(n) \cdot C(m)$.

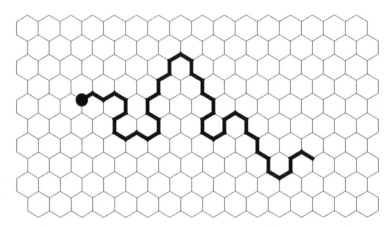

Fig. 6. A self-avoiding walk on the hexagonal lattice. Starting from the origin, we make steps along edges, visiting each vertex at most once. How many such walks of length n are there?

On the other hand, if we glue together two self-avoiding walks of lengths n and m, the resulting walk of length $n + m$ need not be self-avoiding, so that in general there is no equality. This makes the determination of $C(n)$ difficult.

The number μ is called the *connective constant* and has several important applications, so its determination is indeed important. The connective constant μ depends on the grid, which can be established by comparing its values for the hexagonal and square grids in the plane. An IMO-type problem would be to show that

$$\mu_{\text{hex}} < 2 < \mu_{\text{square}}.$$

The inequalities "$\leq$" are easy, can you do it with "$<$"?

Though many estimates were proved, quite some time passed before a guess was even made about the actual values. In 1982, the physicist Bernard Nienhuis found a heuristic argument for the value of μ on the two-dimensional hexagonal lattice, like the one in Figure 6. His argument strongly suggested that in this case

$$C(n) \approx \left(\sqrt{2+\sqrt{2}}\right)^n,$$

and moreover, if one wants higher precision, that for every $\varepsilon > 0$

$$\left(\sqrt{2+\sqrt{2}}\right)^n n^{\frac{11}{32}-\varepsilon} < C(n) < \left(\sqrt{2+\sqrt{2}}\right)^n n^{\frac{11}{32}+\varepsilon},$$

when n is large. His arguments are quite beautiful and inspiring, but they are not mathematically rigorous (and would not have yielded a full score if submitted at a mathematical olympiad).

More than 20 years had to pass before a mathematical solution confirming this prediction was found. Just two months after giving a talk on this subject at the IMO anniversary in Bremen, together with Hugo Duminil-Copin we proved that on a hexagonal lattice indeed $\mu = \sqrt{2+\sqrt{2}}$. Surprisingly, the proof is elementary, and so short that it could be written within the IMO time limit! We carefully count the self-avoiding walks, looking not only at their lengths but also at their windings (i.e. the number of turns they make). The value of μ arises in relation to turns, as $2\cos(\pi/8)$.

So is it surprising that the proof had to wait for so long, and could this problem be given as an IMO problem? The answer to both questions is no: although our way of counting is elementary, inventing it the first time around is far from easy, and requires knowing, in addition to mathematics, a great deal of physics. And what about 11/32? We are probably closer to it than ever before. The mathematicians Greg Lawler, Oded Schramm, and Wendelin Werner have explained where it would come from, and together our results could eventually lead to the proof. But it is still some months or years away, and probably won't be elementary — surprisingly, the number 11/32 arises in a much more complicated way than $2\cos(\pi/8)$!

6 Conclusion

Problem solving experience gained at the IMO will be useful to you regardless of what you decide to do in life. But it will be especially useful if you decide to become a mathematician, and although there are many other things involved in mathematical research besides problem solving, most of them are exciting as well. Mathematics is currently an exciting field, with many beautiful problems and many surprising connections between different branches and to other disciplines. It has become a truly collaborative effort, and is as international as the IMOs — the researchers mentioned in this short article alone come from almost a dozen different countries. I hope that many IMO participants will go on to become mathematicians, and that we will meet again.

Recommended Reading

Among books about the topics discussed, I tried to choose some which are interesting to research mathematicians, and at the same time well written and accessible to motivated high school students. Incidentally, three of the mathematicians mentioned in this article are among the authors of the books below.

There are many books on problem solving, and quite a few apply both to competition and research mathematics. I would like to mention just two (though there are many more that I like).

[1] George Pólya, *How to Solve It*. Princeton University Press, Princeton (1945)
  This is perhaps the first notable book about problem solving, and it has had a great impact. It remains a timeless classic today.
[2] Paul Halmos, *Problems for Mathematicians, Young and Old*. The Dolciani Mathematical Expositions. The Mathematical Association of America, Washington (1991)
  The author has written a few books about problems in research mathematics. This is the most accessible one; it has many problems on the borderline between IMO and research mathematics.

From the many popular books about games, only a few apply to our context: we discuss games without randomness and played by one person only, so that the evolution is completely determined by the initial state. Still, there are some very good ones.

[3] Elwyn R. Berlekamp, John H. Conway, and Richard K. Guy, *Winning Ways for Your Mathematical Plays, second edition*. AK Peters, Wellesley (2004)
  This (very lively) four-volume book discusses the general theory and also describes many examples of non-random games, played by one or several players. Game of Life (which was invented by one of the authors) is discussed in the last chapter.
[4] Joel L. Schiff, *Cellular Automata: A Discrete View of the World*. Wiley-Interscience Series in Discrete Mathematics & Optimization. Wiley-Interscience, Hoboken (2008)
  This is perhaps the best popular introduction to cellular automata, discussing the game of life, the sandpile, and the Ising model, among other things. It is accessible to high-school students while remaining interesting for research mathematicians.

Similarly, there are many popular books about random models of physical phenomena, though most of them stress the physics side of the story.

[5] Gregory F. Lawler and Lester N. Coyle, *Lectures on Contemporary Probability*. Student Mathematical Library, volume 2. The American Mathematical Society, Providence (1999)

This is a short collection of lectures in probability, requiring almost no background. It discusses several very modern research topics, from the Self Avoiding Walk to card shuffling.

[6] Alexei L. Efros, *Physics and Geometry of Disorder: Percolation Theory*. Science for Everyone. Mir, Moscow (1986)

This is an introduction to the domain of mathematics that studies phase transitions by considering random colorings of lattices. It is written very nicely and is specifically oriented towards high school students.

Graph Theory Over 45 Years

László Lovász

Abstract. From 1963 to 1966, when I participated in IMOs, graph theory did not appear in the problem sets. In recent years, however, graph-theoretic problems were often given. What accounts for this? What is the role of graph theory in mathematics today? I will try to answer these questions by describing some of the many connections between graph theory and other areas of mathematics as I encountered them.

1 Introduction

Graph theory is not a new subject. The first result in graph theory was the solution of the Königsberg Bridges Problem in 1736 by Leonhard Euler, one of the greatest mathematicians of all times.

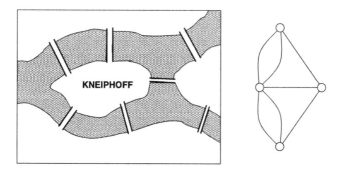

Fig. 1. The bridges of Königsberg in Euler's time, and the graph modeling them.

László Lovász
Institute of Mathematics, Eötvös Loránd University, H-1117 Budapest, Hungary.
e-mail: lovasz@cs.elte.hu

It started with a recreational challenge made by the citizens of Königsberg. The city was divided into four districts by the river Pregel (Figure 1), which were connected by seven bridges. This gave rise to a question: *Is it possible to walk in such a way that each bridge is crossed exactly once?*

Euler proved that such a walk is impossible. If we represent each district by a vertex and draw an edge between two vertices for each bridge that connects the corresponding districts, we get the little graph on the right hand side of Figure 1. The argument of Euler (which is simple and so well known that I will not reproduce it here) can be translated to this graph by examining the degrees of the vertices (that is, the number of edges incident with each vertex).

Many important results in graph theory were obtained in the nineteenth century, mostly in connection with electrical networks. However nobody thought to consider graph theory as an area of mathematics in its own right until the first book on graph theory, Dénes König's *Theorie der endlichen und unendlichen Graphen*, was published in 1936.

König taught in Budapest, and he had two very prominent students, Paul Erdős and Tibor Gallai. Many other Hungarian mathematicians of that generation got interested in graph theory and proved results which are considered fundamental today. Turán and Hajós are two examples.

I was introduced to graph theory, and thereby to mathematical research, quite early in life. My high school classmate and friend Lajos Pósa (himself a gold medalist at the IMO) met Erdős when he was quite young. Erdős gave him graph-theoretic problems to solve, and he was successful. When Pósa started high school, he wrote a paper with Erdős and then wrote more papers by himself. Later, when we met, he told me about other problems due to Erdős, and since I was able to solve one or two, Erdős and I were introduced. Erdős gave me unsolved problems, and thereafter I thought up some of my own, and so began my lifelong commitment to graph theory.

Many of you have probably heard of Erdős. He was not only one of the greatest mathematicians of the twentieth century but a special soul, who did not want to settle, did not want to have property (so that it would not distract him from doing mathematics), and traveled all the time. He was always surrounded by a big group of young people, and he shared with them new problems, research ideas, and results about which he learned during his travels.

Gallai was just the opposite — a very shy and quiet person, who preferred long, one-on-one conversations. When I was a student, I visited him regularly and learned a lot about graph theory and his ideas concerning the directions in which it was developing.

In those days, graph theory was quite isolated from mainstream mathematics. It was often regarded as recreational mathematics, and I was often advised by older mathematicians to do something more serious. I have worked in other areas of mathematics since then (algorithms, geometry, and optimization), but in one way or another there was always some graph-theoretic problem in the background.

Things have changed. Graph theory has become important over the course of the last decades, both through its applications and through its close links with other parts of mathematics. Let me describe some developments.

2 Discrete Optimization

I wrote my thesis under the guidance of Gallai on the problem of factors of graphs (today we call them matchings). The basic question is: *Given a graph, can we pair up its vertices so that the two vertices in each pair are adjacent?* (Such a pairing is called a *perfect matching*.) More generally, what is the maximum number of edges that are pairwise disjoint (that is, without a common endpoint)? A special case is that of bipartite graphs, i.e. graphs in which the vertices are divided into two classes, so that each edge connects two vertices in different classes. In this case, the answer was given by König in 1931: *The maximum number of pairwise disjoint edges in a bipartite graph is equal to the minimum number of vertices covering all edges.* Earlier, in 1914, König proved the following related theorem about bipartite graphs: *The minimum number of colors needed to color the edges of a graph so that edges with a common endpoint are colored differently is equal to the maximum degree of its vertices.*

Tutte extended the characterization of bipartite graphs with perfect matchings to all graphs (the condition is beautiful but a bit too complicated to state here). Many other matching problems remained unsolved, however (allowing me to write a thesis by solving some), and matching theory is still an important source of difficult, but not hopelessly difficult, graph-theoretic problems.

The Austrian mathematician Menger studied the following question in the 1920s: *Given a graph and two vertices s and t, what is the maximum number of mutually disjoint paths from s to t (meaning disjoint except for their common endpoints s and t)?* He proved that this number is equal to the minimum number of vertices (different from s and t) whose removal destroys all paths from s to t. This is a very useful identity, but Menger's proof does not tell us how to compute the aforementioned value. In fact it took thirty years until the American mathematicians Ford and Fulkerson defined flows in networks and used them to give an efficient algorithm for computing a maximum family of mutually disjoint s-t paths in a graph.

The matching problem and the disjoint path problem mentioned above are examples of optimization problems that are quite different from those one studies in calculus. There are many other graph-theoretic optimization problems, some of which, like the Traveling Salesman Problem, became very well known. In a typical optimization problem in analysis, we want to find the minimum or maximum of a function, where the function is "smooth" (differentiable) and defined on an interval. Many of you probably learned how to do this: we find the zeroes of the derivative of the function, and compare the values of the function at these points as well as at the ends of the interval. In discrete optimization, the situation is quite different: we want to optimize functions that are defined on a finite but large and complicated set (like the set of all matchings in a graph, where the function is the number of edges in the matching). These functions have no derivatives, and classical methods of analysis are useless.

There are several methods to attack such problems; perhaps the most successful proceeds by linear programming, which can be thought of as the art of solving systems of linear inequalities. Most of you could probably solve a system of two linear *equations* with two unknowns, or three linear equations with three unknowns. (You could, for example, eliminate one variable by subtracting the equations from each other, and then repeat.) Solving systems of linear *inequalities* is substantially more complicated, but also possible. The methods are analogous, but quite a bit more involved. In most applications, one wants to find not just any solution, but an *optimal* solution, for which a certain linear function is maximized or minimized. There are, however, rather easy methods to reduce this seemingly more involved problem to the first one.

It is perhaps interesting to note that this algebraic problem can be translated to geometry by constructing a convex polyhedron (in a high-dimensional space) and reducing the optimization problem to optimizing a linear function over this polyhedron.

One important source of combinatorial optimization problems are hypergraphs. In an ordinary graph, every edge has two endpoints. We can generalize this, and allow edges that have any number of endpoints. Tibor Gallai called my attention to the fact that any problem in graph theory could be extended (usually in more than one way) to hypergraphs, and that virtually all of these hypergraph problems were unsolved (many of them still are).

For example, König's two theorems mentioned above remain perfectly meaningful for hypergraphs. The question is: how can we define "bipartite" hypergraphs so that the theorems remain not only meaningful but true? A first attempt is to assume that the vertices can be partitioned into two classes, so that every edge meets both classes (it could now contain more than one vertex from a class). We call such hypergraphs 2-*colorable*; they are interesting and important, but in this case, there are easy examples showing that both of König's theorems fail to hold for them. One can try other variations on the notion of bipartiteness, but none of these seem to work. In one of my first papers I managed to prove that the two theorems remain equivalent (even though there is no simple criterion under which they hold). This was a hypergraph-theoretic reformulation of a graph-theoretic conjecture of Berge, and the proof showed that hypergraph theory is useful not only for finding new research problems, but also for solving old problems.

3 Computer Science

Coming back to matching theory, many of us tried to obtain an analogue of Tutte's aforementioned characterization of graphs having a perfect matching for Hamilton cycles: *Given a graph, is there a cycle in it that goes through each vertex exactly once?* The problem is quite similar to the matching problem. It is also quite similar to the Euler cycle problem which started this paper, and which has an easy answer. My advisor Tibor Gallai and many of us were wondering why it was so much more difficult than the other two.

This time (around 1970) was also a time of rapid developments in computer science, in particular of the theory of algorithms and their complexity. In 1972–73, I spent a year in the US and learned about the newly developed theory of polynomial time algorithms and NP-complete problems. The fundamental definitions are as follows.

- A problem is in *class P* if there exists an algorithm that solves it in *polynomial time*, i.e., so that the time required to find a solution is bounded above by a polynomial in the size of the problem input. Such problems are considered "easy", or at least efficiently solvable (no matter what the degree of the polynomial is).
- A problem is in *class NP* (or is an *NP-problem*) if a proposed solution can be verified in polynomial time (whether or not we know a polynomial time algorithm to find the solution). In other words, it can be solved in polynomial time by "lucky guessing", i.e., non-deterministically. The letters NP stand for *non-deterministic polynomial*.
- Finally, a number of NP-problems have the property that if a polynomial-time solution algorithm for one of them was found, it would imply that *all* NP-problems also had polynomial-time solution algorithms (because a polynomial solution to such a problem could be transformed, in polynomial time, to a solution of any other NP problem). Among the NP-problems, these are thus the "hardest" ones; they are known as NP-complete.

It is widely expected that there is a real distinction between class P and class NP: in other words, there are problems for which a solution can be verified but not found in polynomial time. However, no mathematical proof is known. This fundamental question of complexity theory, usually phrased "P = NP ?", was named one of the seven most important open problems in mathematics in 2000.

The theory of the complexity of algorithms thrilled me because it explained the difference between the matching problem and the Hamilton cycle problem: the first one was in P, and the other one was NP-complete!

When I returned to Hungary, I met a friend of mine, Péter Gács, who had spent a year in Moscow. Interrupting one another, we began to describe the great ideas we had learned about: Leonid Levin's work in Moscow, and the work of Cook and Karp in the US. As it turned out, they were independent developments of the same theory. (For about two weeks we thought we had a proof of the P ≠ NP conjecture. Nowadays, we would be more suspicious of simplistic ideas concerning a famous problem....)

Graph theory has become one of the most prominent areas of the mathematical foundation of computer science. We have seen that graph-theoretic problems motivated the "P = NP?" problem and many more of the most interesting questions that arose in the development of complexity theory.

There is also an important connection that points in the other direction: to describe the process of a complicated computation in mathematical terms, one needs the notion of a directed graph. Steps in the computation are represented by vertices (often called "gates"), and an edge indicates that the output of one step is the input of the other. We can assume that these outputs are just bits, and the gates themselves can be very simple (it suffices to use just one kind, a NAND gate, which outputs TRUE if and only if at least one of the inputs is FALSE). All of the complexity of the computation goes into the structure of the graph.

I am sorry to report that we graph theorists have not achieved much in this direction. For example, the famous "P = NP?" problem boils down to the following question: we want to design a network that can find out whether an arbitrary graph with n vertices has a Hamilton cycle or not. The vertices of the graph are labeled $1, 2, 3, \ldots, n$. The network has $\binom{n}{2}$ input gates $v_{i,j}$ ($1 \leq i < j \leq n$) and a single output gate u. The graph is specified by assigning TRUE to an input gate $v_{i,j}$ if and only if the vertices i and j are connected by an edge; the other input gates are assigned FALSE. We want the output to be TRUE if and only if the graph has a Hamilton cycle, for all possible input graphs. Such a network can be designed, but the question is whether its size can be bounded by some polynomial in n, say by n^{100}. To understand the complexity of computations using graph theory is a BIGGGG challenge!

4 Probability

Around 1960, Paul Erdős and Alfréd Rényi developed the theory of random graphs. In their model, we start with n vertices, which we fix. Then we begin to add edges in such a way that new edges are chosen randomly among all pairs of vertices that are not yet adjacent. After adding a prescribed number m of edges, we stop.

Of course, if we repeat this construction, we will very likely end up with a different graph. If n and m are large, however, the graphs constructed in this way will be very similar, with a very small probability of getting an "outlier" (this is a manifestation of the Law of Large Numbers). A related phenomenon is that in watching these random graphs develop (as edges are added), we observe sudden changes in their structure. For example, if we look at the graph when it has $m = 0.49n$ edges, it will almost surely consist of many small connected components. If we look at it again when it has $0.51n$ edges, then it will contain a single giant component (containing about 4% of all vertices, independently of n), along with a few very small ones (the sizes of which are small compared to n). This sudden change in the structure of the graph is closely related to everyday physical phenomena like the melting of ice. This is called a "phase transition", and its study is a very active (and quite difficult) area.

Determining typical properties of random graphs is not easy, but Erdős and Rényi worked out many. Less than a decade later, I learned probability theory from the lectures of Rényi, and he gave me copies of their papers on random graphs. I have to admit that I was not interested in them for a while. They contained long, detailed computations, and who likes to read such things? Since then, the field has blossomed into one of the most active areas in graph theory and has become fundamental for modeling the internet. Of course I could not avoid working with random graphs, as we shall see below.

Probability enters graph theory in other ways. In fact, it is becoming a fundamental tool in many areas of mathematics. Often questions may have nothing to do with probability, although their solutions involve random choice. Their proofs can be so simple and elegant that I can describe one of them here. Let H be a hypergraph. Under what conditions is it 2-colorable? For an ordinary graph, this is a classical question that can be answered easily (one possible answer is that the graph contain no odd cycles), but for a general hypergraph, this is a very hard question.

Erdős and Hajnal proved the following theorem around 1970: *If every edge of the hypergraph H has r vertices, and H has less than 2^{r-1} edges, then H is 2-colorable.* This is an example of a hypergraph question that "stands alone": the claim is trivial for graphs (you need 3 edges to create a non-bipartite graph), but very interesting for general hypergraphs.

Trying to prove this via usual methods (e.g. induction) does not work. But here is a proof using probability theory. Let us color the vertices at random: for each vertex, we flip a coin and color it red or blue depending on the outcome of the coin flip. There are possible "bad events": the event that a given edge has only one color is bad. If we are lucky, and none of these occur, then we get a good coloring. But can all bad events be avoided simultaneously? What is the probability of this?

Let us start with an easier task: what is the probability that a specific bad event is avoided? There are 2^r ways to color the vertices of an edge (the colors of the other vertices don't matter), and two of these are bad. So the probability that this edge is all red or all blue is $2/2^r = 2^{1-r}$.

Now the probability that one of these bad events happens is bounded by the sum of their probabilities. This is less than $2^{r-1} \times 2^{1-r} = 1$. Thus, good colorings must exist. It is difficult to construct a single good coloring, but coloring the vertices randomly works!

This method, called the *probabilistic method*, has become very powerful and important. I got involved in improving it. For example, I proved (working with Erdős) that in the above problem, we don't have to limit the number of edges; it suffices to assume that no edge meets more than 2^{r-3} other edges.

5 Algebra, Topology, and Graph Theory

Probability is not the only field of mathematics that has profound applications in graph theory. There are beautiful applications of very classical mathematics, such as algebra or topology. I always found this fascinating, and tried to find connections myself. When I was a student, it seemed that mathematics was on a path towards fragmentation: different branches, and in particular relatively new branches like probability or graph theory, looked to be separating both from each other and from classical branches. I am happy to report, however, that this tendency seems to have turned around, and that the case for the unity of mathematics is on much firmer grounds.

This may sound like empty speculation, but I can illustrate such connections by citing a recent IMO problem, namely Problem 6 of 2007, and its solution. The problem was the following:

> Let n be a positive integer, and consider the set
> $$S = \{(x, y, z) \in \{0, \ldots, n\}^3 \mid x + y + z > 0\}$$
> as a set of $(n+1)^3 - 1$ points in three-dimensional space. Determine the smallest number of planes whose union contains S but does not contain the point $(0, 0, 0)$.

It is easy to guess the right answer (it is $3n$) and to find a collection of $3n$ planes that satisfy the given conditions. In order to prove that $3n$ is really minimal, it seems natural to try combinatorial approaches like induction, colorings, or graph theory. But although the problem is apparently a combinatorial one, it turns out that the only known way to get to a solution is to apply algebraic methods — for instance, the theory of polynomials. Suppose you can find $m < 3n$ planes that cover S but don't pass through the origin. They are described by m linear equations in three variables x, y, and z, as follows:

$$a_i x + b_i y + c_i z + d_i = 0 \qquad (d_i \neq 0, \quad 1 \leq i \leq m).$$

Multiplying these equations together, we obtain the polynomial

$$P(x, y, z) = \prod_{i=1}^{m} (a_i x + b_i y + c_i z + d_i)$$

of degree m that is zero on S but nonzero at the origin.

Now we define an operator Δ_x that replaces a polynomial Q with the polynomial $\Delta_x Q$, where

$$\Delta_x Q(x,y,z) = Q(x+1,y,z) - Q(x,y,z),$$

and analogous operators Δ_y and Δ_z (these are discrete versions of differentiation operators). Note that each of these operators decreases the degree of a polynomial by at least 1 (we define the zero polynomial to have degree $-\infty$).

By repeatedly applying these three operators to our polynomial P, one sees inductively that $\Delta_x^r \Delta_y^s \Delta_z^t P(0,0,0)$ is never zero for $r,s,t \leq n$, and in particular that $\Delta_x^n \Delta_y^n \Delta_z^n P(0,0,0) \neq 0$. But this contradicts the fact that the polynomial $\Delta_x^n \Delta_y^n \Delta_z^n P$ has degree at most $m - 3n < 0$, so it has to be the zero polynomial.

This very difficult IMO problem was solved by only five students, and all of them gave solutions very similar to the one described above. (This solution is due to Peter Scholze, one of these five students.)

6 Networks, or Very Large Graphs

Many areas of mathematics, computer science, biology, physics, and the social sciences are concerned with properties of very large graphs (often called networks).

The internet is an obvious example. There is in fact more than one network that we can define based on the internet. One is a "physical" network, i.e. the graph whose vertices are electronic devices (computers, telephones, routers, hubs, etc.), and whose edges are the connections between them (wired or wireless). There is also a "logical" network, often called the World Wide Web, whose vertices are the documents available on the internet, and whose (directed) edges are the hyperlinks that point from one to the other.

Social networks are of course formed by people, and they may be based on various definitions of connectivity. However the best known and best studied social networks are internet-based (like Facebook). Some historians want to understand history based on a network of humans. The structure of this network determines, among other things, how fast news, disease, religion, and knowledge spread through society, and it has an enormous impact on the course of history.

There are many other networks related to humans. The brain is a great example of a huge network whose workings are not yet fully understood. It is too large for its structure (i.e. all of its neurons and their connections) to be encoded in our DNA. Why is it that it still functions and is able, for example, to solve math problems?

Biology is full of systems whose basic structure is a network. Consider the interactions between the plants and animals living in a forest (who eats whom?), or the interactions between proteins in our bodies. Networks are about to become part of a basic language for describing the systems and structures in many parts of nature — just as continuous functions and the operations of differentiation and integration are part of a basic language for describing mechanics and electromagnetism.

From the point of view of a mathematician, this should imply that powerful tools must be created to help biologists, historians, and sociologist describe the systems in which they (and all of us) are interested. This will not be an easy task, since these systems are very diverse. Modeling traffic, information distribution, and the electrical networks discussed above is only the tip of the iceberg.

Let me conclude with a few words about a topic that I have studied recently which is motivated by problems concerning very large graphs. The main idea is to assume that these graphs "tend to infinity" and to study the "limit objects". We often use the finite to approximate the infinite; obtaining numerical solutions to physical equations (say, for the purpose of predicting the weather) usually requires restricting space and time to a finite number of points, and then computing (more or less step-by-step) how temperature, pressure, etc. develop at these points.

The idea that the infinite may be a good approximation of the finite is more subtle. Continuous structures are often cleaner, more symmetric, and richer than their discrete counterparts.

To illustrate this idea with a physical example, consider a large piece of metal. This piece of metal is a crystal that is really a large graph consisting of atoms and bonds between them (arranged in a periodic and therefore rather boring way). But for an engineer who uses this metal to build a bridge, it is more useful to consider it as a continuum with a few important parameters (e.g. density, elasticity, and temperature) that are functions on this continuum. Our engineer can then use differential equations to compute the stability of the bridge. Can we regard a very large graph as some kind of a continuum?

In some cases this is possible, and Figure 2 illustrates the idea. We start with a random graph that is just a little more complicated than the random graphs introduced by Erdős and Rényi. It is constructed randomly according to the following rule: at each step, either a new vertex or a new edge is created. If the total number of vertices is n, then the probability that a new vertex is created is $1/n$, and the probability that a new edge is created is $(n-1)/n$. A new edge connects a randomly chosen pair of vertices.

The grid on the left represents a random graph with 100 vertices in the following way: the pixel at the intersection of the i-th row and the j-th column is black if there is an edge connecting the i-th and j-th vertices, and white if there is no such edge. Thus the area in the upper left is darker, because a pixel there represents a pair of vertices that have been around for a longer time and hence have a greater chance of being connected.

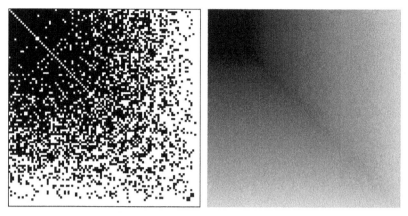

Fig. 2. A randomly grown uniform attachment graph with 100 vertices, and the continuous function $1 - \max(x, y)$ approximating it.

Although this graph is random, the pixel picture on the left is, from a distance, quite similar to the continuous function $1 - \max(x, y)$, which is depicted on the right. If instead of 100 vertices we took 1000, the similarity would be even more striking. One can prove that the rather simple function on the right encodes all of the information one needs to know about the graph on the left, except for random fluctuations, which become smaller and smaller as the number of vertices grows.

Large graphs, with thousands of vertices, and huge graphs, with billions, represent a new kind of challenge for the graph theorists. It hints at the beauty of mathematics that to meet some of these challenges we have to discover and use more and more connections with other, more classical parts of mathematics.

Further Reading

[1] J. Adrian Bondy and Uppaluri S. R. Murty, *Graph Theory.* Graduate Texts in Mathematics, volume 244. Springer, New York (2008)
[2] László Lovász, József Pelikán, and Katalin Vesztergombi, *Discrete Mathematics: Elementary and Beyond.* Springer, New York (2003)

Communication Complexity

Alexander A. Razborov

Abstract. When I was asked to write a contribution for this book about something related to my research, I immediately thought of communication complexity. This relatively simple but extremely beautiful and important sub-area of complexity theory studies the amount of *communication* needed for several distributed parties to learn something new. We will review the basic communication model and some of the classical results known for it, sometimes even with proofs. Then we will consider a variant in which the players are allowed to flip fair unbiased coins. We will finish with a brief review of more sophisticated models in which our current state of knowledge is less than satisfactory. All our definitions, statements and proofs are completely elementary, and yet we will state several open problems that have evaded strong researchers for decades.

1 Introduction

As the reader can guess from the name, communication complexity studies ways to arrange *communication* between several parties so that at the end of the day they learn what they are supposed to learn, and to do this in the most efficient, or least *complex* way. This theory constitutes a small but, as we will see below, very beautiful and important part of *complexity theory* which, in turn, is situated right at the intersection of mathematics and theoretical computer science. For this reason I would like to begin with a few words about complexity theory in general and what kind of problems researchers

---

Alexander A. Razborov
Department of Computer Science, The University of Chicago, 1100 East 58th Street, Chicago, IL 60637, USA; and Steklov Mathematical Institute, Moscow, Russia, 117 966. e-mail: razborov@cs.uchicago.edu

are studying there. The reader favoring concrete mathematical content over philosophy should feel free to skip the introduction and go directly to Section 2.

Complexity theory is interested in problems that in most cases can roughly be described as follows. Assume that we have a task T that we want to accomplish. In most cases this involves one or more computers doing something, but it is not absolutely necessary. There can be many different ways to achieve our goal, and we denote the whole set of possibilities by $\mathcal{P}_T$. Depending on the context, elements of $\mathcal{P}_T$ can be called *algorithms* or, as in our article, *protocols*. In most cases of interest it is trivial that there is at least one algorithm/protocol to solve T, that is the set $\mathcal{P}_T$ is non-empty.

While all $P \in \mathcal{P}_T$ solve our original task T, not all solutions are born equal. Some of them may be better than others because they are shorter, consume less resources, are simpler, or for any other reason. The main idea of mathematical complexity theory is to try to capture our intuitive preferences by a positive real-valued function $\mu(P)$ ($P \in \mathcal{P}_T$) called *complexity measure* with the idea that the smaller $\mu(P)$ is, the better is our solution P. Ideally, we would like to find the *best* solution $P \in \mathcal{P}_T$, that is the one minimizing the function $\mu(P)$. This is usually very hard to do, so in most cases researchers try to approach this ideal from two opposite sides as follows:

- try to find "reasonably good" solutions $P \in \mathcal{P}_T$ for which $\mu(P)$ may perhaps not be minimal, but still is "small enough". Results of this sort are called "upper bounds" as what we are trying to do mathematically is to prove *upper bounds* on the quantity

$$\min_{P \in \mathcal{P}_T} \mu(P)$$

that (not surprisingly!) is called the *complexity* of the task T.
- *Lower bound* problems: for some $a \in \mathbb{R}$ we try to show that $\mu(P) \geq a$ for any P, that is that there is no solution in $\mathcal{P}_T$ better than a. The class $\mathcal{P}_T$ is usually very rich and solutions $P \in \mathcal{P}_T$ can be based upon very different and often unexpected ideas. We have to take care of all of them with a uniform argument. This is why lower bound problems are amongst the most difficult in modern mathematics, and the overwhelming majority of them is still wide open.

All right, it is a good time for some examples. A great deal of mathematical olympiad problems are actually of complexity flavor even if it is not immediately clear from their statements.

Communication Complexity 99

> You have 7 (or 2010, n, ...) coins, of which 3 (at most 100, an unknown number, ...) are counterfeit and are heavier (are lighter, weigh one ounce more, ...) than the others. You also have a scale that can weigh (compare, ...) as many coins as you like (compare at most 10 coins, ...). How many weighings do you need to identify all (one counterfeit, ...) coin(s)?

These are typical complexity problems, and they are very much related to what is called *sorting networks and algorithms* in the literature. The task T is to identify counterfeit coins, and $\mathcal{P}_T$ consists of all sequences of weighings that allow us to accomplish it. The complexity measure $\mu(P)$ is just the length of P (that is, the number of weighings used).

> You have a number (polynomial, expression, ...), how many additions (multiplications, ...) do you need to build it from certain primitive expressions?

Not only is this a complexity problem, but also a paradigmatic one. Can you describe $T, \mathcal{P}_T$ and μ in this case? And, by the way, if you think that the "school" method of multiplying integers is optimal in terms of the number of digit operations used, then this is incorrect. It was repeatedly improved in the work of Karatsuba [13], Toom and Cook (1966), Schönhage and Strassen [26] and Fürer [11], and it is still open whether Fürer's algorithm is the optimal one. It should be noted, however, that these advanced algorithms become more efficient than the "school" algorithm only for rather large numbers (typically at least several thousand digits long).

If you have heard of the famous **P** vs. **NP** question (otherwise I recommend to check out e.g. http://www.claymath.org/millennium/P_vs_NP), it is another complexity problem. Here T is the task of solving a fixed **NP**-complete problem, e.g. SATISFIABILITY, and $\mathcal{P}_T$ is the class of all deterministic algorithms fulfilling this task.

In this article we will discuss complexity problems involving *communication*. The model is very clean and easy to explain, but quite soon we will plunge into extremely interesting questions that have been open for decades... And, even if we will not have time to discuss it here at length, the ideas and methods of communication complexity penetrate today virtually all other branches of complexity theory.

Almost all material contained in our article (and a lot more) can be found in the classical book [17]. The recent textbook [3] on computational complexity has Chapter 13 devoted entirely to communication complexity, and you can find its applications in many other places all over the book.

Since the text involves quite a bit of notation, some of it is collected together at the end of the article, along with a brief description.

2 The Basic Model

The basic (deterministic) model was introduced in the seminal paper by Yao [27]. We have two players traditionally (cf. the remark below) called Alice and Bob, we have finite sets X, Y and we have a function $f : X \times Y \longrightarrow \{0,1\}$. The task T_f facing Alice and Bob is to evaluate $f(x,y)$ for a given input (x,y). The complication that makes things interesting is that Alice holds the first part $x \in X$ of their shared input, while Bob holds another part $y \in Y$. They do have a two-sided communication channel, but it is something like a transatlantic phone line or a beam communicator with a spacecraft orbiting Mars. Communication is expensive, and Alice and Bob are trying to minimize the number of bits exchanged while computing $f(x,y)$.

Thus, a protocol $P \in \mathcal{P}_T$ looks like this (see Figure 1). Alice sends a

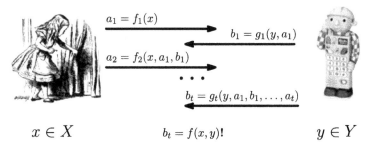

Fig. 1. Protocol P for computing $f(x,y)$. (Picture of Alice by John Tenniel.)

message encoded for simplicity as a *binary string* a_1 (i.e., a finite sequence of zeroes and ones). Bob responds with some b_1 that depends only on his y *and* Alice's message a_1. They continue in this way until one of them (say, Bob) is able to compute the value of $f(x,y)$ and communicate it to Alice in the t-th round.

Remark 1. It should be noted that Alice and Bob are by far the most lovable and popular heroes in the whole literature on complexity and the closely related field of cryptography. As such, they are summoned up in many other episodes of this ongoing story, and, just like their real-life prototypes, they live a busy life. Sometimes their goals coincide only partially, and they are very cautious about leaking out unwanted information, then it is called cryptography. Often there is an evil eavesdropper (usually called, for obvious reasons, Eve). Sometimes Alice and Bob do not even trust that the other party will truthfully follow the protocol, although in such cases they usually change their names to Arthur and Merlin. But in our article we will consider only the simplest scenario: complete mutual trust, nothing to hide, perfect and secure communication channel.

In this definition we deliberately left a few things imprecise. For example, is the *length* of Alice's message a_1 fixed or is it allowed to depend on x? Likewise, can the number of rounds t depend on x and y and, if so, how can Alice know that Bob's message b_t is actually the last one and already gives the final answer? It turns out, however, that all these details are very inessential, and the reader can fill them any way he or she likes — this will change the complexity only by a small additive factor.

How to measure the complexity $\mu(P)$ of this protocol P? There are several ways of doing this, all of them reasonable. In this article we will focus only on the most important and popular model called *worst-case complexity*. For any *given* input $(x,y) \in X \times Y$ we define the *cost* of the protocol P on this input as the total number of bits[1] $|a_1|+|b_1|+\ldots+|b_t|$ exchanged on this input (cf. Figure 1). And then we define the complexity (that, for historical reasons, is also called *cost* in this case) cost(P) of the protocol P as the *maximal* cost of P over all inputs $(x,y) \in X \times Y$. Finally, the *communication complexity* $C(f)$ of (computing) the function $f: X \times Y \longrightarrow \{0,1\}$ is defined as the *minimum* $\min_{P \in \mathcal{P}_f}$ cost(P) taken over all legitimate protocols P, i.e., those protocols that correctly output the value $f(x,y)$ for all possible inputs. We would like to be able to compute $C(f)$ for "interesting" functions f, or at least get good estimates for it.

The first obvious remark is that

$$C(f) \leq \lceil \log_2 |X| \rceil + 1 \qquad (1)$$

for any problem[2] f. The protocol of this cost is very simple: Alice encodes her input x as a binary string of length $\lceil \log_2 |X| \rceil$ using any injective encoding $f_1: X \longrightarrow \{0,1\}^{\lceil \log_2 |X| \rceil}$ and sends $a_1 = f_1(x)$ to Bob. Then Bob decodes the message (we assume that the encoding scheme f_1 is known to both parties in advance!) and sends the answer $f(f_1^{-1}(a_1), y)$ back to Alice.

Surprisingly, there are only very few interesting functions f for which we can do significantly better than (1) in the basic model. One example that is sort of trivial is this. Assume that X and Y consist of integers not exceeding some fixed N: $X = Y = \{1,2,\ldots,N\}$. Alice and Bob want to compute the $\{0,1\}$-valued function $f_N(x,y)$ that outputs 1 if and only if $x+y$ is divisible by 2010. A much more economical way to solve this problem would be for Alice to send to Bob not her whole input x, but only its remainder x mod 2010. Clearly, this still will be sufficient for Bob to compute $x+y$ mod 2010 (and hence also $f_N(x,y)$), and the cost of this protocol is only $\lceil \log_2 2010 \rceil + 1 \, (=12)$. Thus,

$$C(f_N) \leq \lceil \log_2 2010 \rceil + 1 \, . \qquad (2)$$

---

[1] $|a|$ is the length of the binary word a.
[2] Note that complexity theorists often identify functions f with computational problems they naturally represent. For example, the equality *function* EQ_N defined below is also viewed as the *problem* of checking if two given strings are equal.

Now, complexity theorists are lazy people, and not very good at elementary arithmetic. What is really remarkable about the right-hand side of (2) is that it represents *some* absolute constant that magically does not depend on the input size at all! Thus, instead of calculating this expression, we prefer to stress this fact using the mathematical *big-O* notation and write (2) in the simpler, even if weaker, form

$$C(f_N) \leq O(1).$$

This means that there exists a positive universal constant $K > 0$ that anyone interested can (usually) extract from the proof such that for all N we have $C(f_N) \leq K \cdot 1 = K$. Likewise, $C(f_N) \leq O(\log_2 N)$ would mean that $C(f_N) \leq K \log_2 N$ etc. We will extensively use this standard[3] notation in our article.

Let us now consider a simpler problem that looks as fundamental as it can only be. We assume that $X = Y$ are equal sets of cardinality N. The reader may assume that this set is again $\{1, 2, \ldots, N\}$, but now this is not important. The *equality function* EQ_N is defined by letting $\text{EQ}_N(x, y) = 1$ if and only if $x = y$. In other words, Alice and Bob want to check if their files, databases etc. are equal, which is clearly an extremely important task in many applications.

We can of course apply the trivial bound (1), that is, Alice can simply transmit her whole input x to Bob. But can we save even a little bit over this trivial protocol? At this point I would like to strongly recommend you to put this book aside for a while and try out a few ideas toward this goal. That would really help to better appreciate what will follow.

---

[3] We should warn the reader that in most texts this notation is used with the equality, rather than inequality, sign, i.e., $C(f_N) = O(\log_2 N)$ in the previous example. However, we see numerous issues with this usage and in particular it becomes rather awkward and uninformative in complicated cases.

3 Lower Bounds

Did you have any luck? Well, you do not have to be distressed by the result since it turns out that the bound (1) actually can *not* be improved, that is *any* protocol for EQ_N must have cost at least $\log_2 N$. This was proven in the same seminal paper by Yao [27], and many ideas from that paper determined the development of complexity theory for several decades to follow. Let us see how the proof goes, the argument is not very difficult but it is very instructive.

We are given a protocol P of the form shown on Figure 1, and we know that upon executing this protocol Bob knows $\mathrm{EQ}_N(x,y)$. We should somehow conclude that $\mathrm{cost}(P) \geq \log_2 N$.

One very common mistake often made by new players in the lower bounds game is that they begin telling P what it "ought to do", that is, consciously or unconsciously, begin making assumptions about the best protocol P based on the good common sense. In our situation a typical argument would start off by something like "let i be the first bit in the binary representation of x and y that the protocol P compares". "Arguments" like this are dead wrong since it is not clear at all that the best protocol should proceed in this way, or, to that end, in any other way we would consider "intelligent". Complexity theory is full of ingenious algorithms and protocols that do something strange and apparently irrelevant almost all the way down, and only at the end of the day they conjure the required answer like a rabbit from the hat — we will see one good example below. The beauty and the curse of complexity theory is that we should take care of all protocols with seemingly irrational (in our opinion) behavior all the same, and in our particular case we may not assume *anything* about the protocol P besides what is explicitly shown on Figure 1.

Equipped with this word of warning, let us follow Yao and see what useful information we still can retrieve from Figure 1 alone. Note that although we are currently interested in the case $f = \mathrm{EQ}_N$, Yao's argument is more general and can be applied to any function f. Thus, for the time being we assume that f is an arbitrary function whose communication complexity we want to estimate; we will return to EQ_N in Corollary 2.

The first thing to do is to introduce an extremely useful concept of a *history* or a *transcript*: this is the whole sequence $(a_1, b_1, \ldots, a_t, b_t)$ of messages exchanged by Alice and Bob during the execution of the protocol on some particular input. This notion is very broad and general and is successfully applied in many different situations, not only in communication complexity.

Next, we can observe that there are at most $2^{\mathrm{cost}(P)}$ different histories as there are only that many different strings[4] of length $\mathrm{cost}(P)$. Given any fixed history h, we can form the set R_h of all those inputs (x,y) that lead to this history. Let us see what we can say about these sets.

---

[4] Depending on finer details of the model, histories may have different length, the placement of commas can be also important etc. that might result in a slight increase of this number. But remember that we are lazy and prefer to ignore small additive, or even multiplicative factors.

First of all, every input (x,y) leads to one and only one history. This means that the collection $\{R_h\}$ forms a *partition* or *disjoint covering* of the set of all inputs $X \times Y$:
$$X \times Y = \dot{\bigcup}_{h \in \mathcal{H}} R_h, \tag{3}$$
where $\mathcal{H}$ is the set of all possible histories. The notation $\dot{\bigcup}$ stands for *disjoint union* and simultaneously means two different things: that $X \times Y = \bigcup_{h \in \mathcal{H}} R_h$, and that $R_h \cap R_{h'} = \emptyset$ for any two different histories $h \ne h' \in \mathcal{H}$.

Now, every history h includes the value of the function $f(x,y)$ as Bob's last message b_t. That is, any R_h is an f-*monochromatic* set, which means that either $f(x,y) = 0$ for all $(x,y) \in R_h$ or $f(x,y) = 1$ for all such (x,y).

Finally, and this is very crucial, every R_h is a *combinatorial rectangle* (or simply a rectangle), that is it has the form $R_h = X_h \times Y_h$ for some $X_h \subseteq X$, $Y_h \subseteq Y$. In order to understand why, we should simply expand the sentence "(x,y) leads to the history $(a_1, b_1, \ldots, a_t, b_t)$". Looking again at Figure 1, we see that this is equivalent to the set of "constraints" on (x,y) shown there: $f_1(x) = a_1$, $g_1(y, a_1) = b_1$, $f_2(x, a_1, b_1) = a_2, \ldots, g_t(y, a_1, \ldots, a_t) = b_t$. Let us observe that odd-numbered constraints in this chain depend only on x (remember that h is fixed!); let us denote by X_h the set of those $x \in X$ that satisfy all these constraints. Likewise, let Y_h be the set of all $y \in Y$ satisfying even-numbered constraints. Then it is easy to see that we precisely have $R_h = X_h \times Y_h$!

Let us summarize a little bit. For any protocol P solving our problem $f : X \times Y \longrightarrow \{0,1\}$, we have been able to chop $X \times Y$ into at most $2^{\text{cost}(P)}$ pieces so that each such piece is an f-monochromatic combinatorial rectangle. Rephrasing somewhat, let us denote by $\chi(f)$ (yes, complexity theorists love to introduce complexity measures!) the minimal number of f-monochromatic rectangles into which we can partition $X \times Y$. We thus have proved, up to a small multiplicative constant that may depend on finer details of the model:

Theorem 1 (Yao). $C(f) \ge \log_2 \chi(f)$. □

Let us return to our particular case $f = \text{EQ}_N$. All f-monochromatic combinatorial rectangles can be classified into 0-*rectangles* (i.e., those on which f is identically 0) and 1-*rectangles*. The function EQ_N has many large 0-rectangles. (Can you find one?) But all its 1-rectangles are very primitive, namely every such rectangle consists of just one point (x,x). Therefore, in order to cover even the "diagonal" points $\{(x,x) \mid x \in X\}$, one needs N different 1-rectangles, which proves $\chi(\text{EQ}_N) \ge N$. Combining this with Theorem 1, we get the result we were looking for:

Corollary 2. $C(\text{EQ}_N) \ge \log_2 N$. □

Exercise 1. *The function* LE_N *(less-or-equal) is defined on* $\{1, 2, \ldots, N\} \times \{1, 2, \ldots, N\}$ *as*
$$\text{LE}_N(x,y) = 1 \text{ iff } x \le y.$$
Prove that $C(\text{LE}_N) \ge \log_2 N$.

Exercise 2 (difficult). *The function* DISJ_n *is defined on* $\{0,1\}^n \times \{0,1\}^n$ *as*

$$\text{DISJ}_n(x,y) = 1 \text{ iff } \forall i \leq n : x_i = 0 \vee y_i = 0 ,$$

that is, the sets of positions where the strings x and y have a 1 are disjoint. Prove that $C(\text{DISJ}_n) \geq \Omega(n)$.

(Here Ω is yet another notation that complexity theorists love. It is dual to "big-O"; the inequality $C(\text{DISJ}_n) \geq \Omega(n)$ means that there exists a constant $\varepsilon > 0$ that we do not want to compute such that $C(\text{DISJ}_n) \geq \varepsilon n$ for all n.)

Hint. How many points (x,y) with $\text{DISJ}_n(x,y) = 1$ do we have? And what is the maximal size of a 1-rectangle?

4 Are These Bounds Tight?

The next interesting question is, how good is Theorem 1 in general? Can it be the case that $\chi(f)$ is small, that is we do have a good disjoint covering by f-monochromatic rectangles, and nonetheless $C(f)$ is large, so that in particular we can not convert our covering into a decent communication protocol? Figure 2 suggests at least that this question may be non-trivial: it

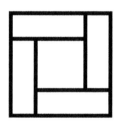

Fig. 2. What should Alice do?

gives an example of a disjoint covering by only five rectangles that does not correspond to any communication protocol.

As in many similar situations, the answer depends on how precise you want it to be. In the next influential paper on communication complexity [1], the following was proved among other things:

Theorem 3 (Aho, Ullman, Yannakakis). $C(f) \leq O(\log_2 \chi(f))^2$.

The proof is not very difficult, but still highly non-trivial. The reader can try to find it by himself or consult e.g. [17].

Can we remove the square in Theorem 3? For almost thirty years that have elapsed since the paper [1], many people have tried to resolve the question one or the other way. But it has resisted all efforts so far...

Open Problem 1. *Is it true that $C(f) \leq O(\log_2 \chi(f))$?*

Besides Theorem 3, the paper [1] contains many other great things pertaining to the so-called *non-deterministic communication complexity*. In this model, Alice and Bob are also given access to a shared string z not determined by the protocol (whence comes the name) but rather given to them by a third all-powerful party trying to convince them that $f(x,y) = 1$. We require that a convincing string z exists if and only if $f(x,y)$ is *indeed* equal to 1, and we note that in this definition we give up on the symmetry of answers 0 and 1. Due to lack of space we discuss this important concept only very briefly, and complexity measures we mention during the discussion will hardly be used in the rest of the article.

Define $t(f)$ in the same way as $\chi(f)$, only now we allow the monochromatic rectangles in our cover to overlap with each other. Clearly, $t(f) \leq \chi(f)$, but it turns out that the bound of Theorem 3 still holds: $C(f) \leq O(\log_2 t(f))^2$. On the other hand, there are examples for which $C(f)$ is of order $(\log_2 t(f))^2$. This means that the (negative) solution to the analogue of Problem 1 for not necessarily disjoint coverings is known.

Let $\chi_0(f)$ and $\chi_1(f)$ be defined similarly to $\chi(f)$, except now we are interested in a disjoint rectangular covering of only those inputs that yield value 0 (respectively, value 1); note that $\chi(f) = \chi_0(f) + \chi_1(f)$. Then still $C(f) \leq O(\log_2 \chi_1(f))^2$ and (by symmetry) $C(f) \leq O(\log_2 \chi_0(f))^2$. By analogy, we can also define the quantities $t_0(f)$ and $t_1(f)$ (the non-deterministic communication complexity we mentioned above turns out to be equal to $\log_2 t_1(f)$). We cannot get any reasonable (say, better than exponential) bound on $C(f)$ in terms of $\log_2 t_1(f)$ or $\log_2 t_0(f)$ only: for example, $t_0(\text{EQ}_N) \leq O(\log_2 N)$ (why?) while, as we already know, $C(\text{EQ}_N) \geq \log_2 N$. In conclusion, there is no good bound on the deterministic communication complexity in terms of the non-deterministic one, but such a bound becomes possible if we know that the non-deterministic communication complexity of the negated function is also small.

The next landmark paper we want to discuss is the paper [19] that introduced to the area *algebraic methods*. So far all our methods for estimating $\chi(f)$ from below (Corollary 2 and Exercises 1 and 2) were based on the same unsophisticated idea: select "many" inputs $D \subseteq X \times Y$ such that every f-monochromatic rectangle R may cover only "a few" of them, and then apply the pigeonhole principle. This method does not use anyhow that the covering (3) is disjoint or, in other words, it can be equally well applied to bounding from below $t(f)$ as well as $\chi(f)$. Is it good or bad? The answer depends. It is always nice, of course, to be able to prove more results, like lower bounds on the *non-deterministic* communication complexity $\log_2 t_1(f)$, with the same shot. But sometimes it turns out that the quantity analogous to $t(f)$ is *always* small and, thus, if we still want to bound $\chi(f)$ from below, we must use methods that "feel" the difference between these two concepts. The *rank lower bound* of Mehlhorn and Schmidt [19] was the first of such methods.

We will need the most basic concepts from linear algebra like a *matrix* M or its *rank* $\mathrm{rk}(M)$, as well as their simplest properties. If the reader is not yet familiar with them, then this is a perfect opportunity to grab any textbook in basic linear algebra and read a couple of chapters from it. You will have to learn this eventually anyway, but now you will also immediately see quite an unexpected and interesting application of these abstract things.

Given any function $f : X \times Y \longrightarrow \{0, 1\}$, we can arrange its values in the form of the *communication matrix* M_f. Rows of this matrix are enumerated by elements of X, its columns are enumerated by elements of Y (the order is unimportant in both cases), and in the intersection of the x-th row and the y-th column we write $f(x, y)$. The following result relates two quite different worlds, those of combinatorics and linear algebra.

Theorem 4. $\chi(f) \geq \mathrm{rk}(M_f)$.

Proof. The proof is remarkably simple. Let $R_1, \ldots, R_\chi$ be disjoint 1-rectangles covering all (x, y) with $f(x, y) = 1$ so that $\chi \leq \chi(f)$. Let $f_i : X \times Y \longrightarrow \{0, 1\}$ be the *characteristic* function of the rectangle R_i, i.e., $f_i(x, y) = 1$ if and only if $(x, y) \in R_i$, and let $M_i = M_{f_i}$ be its communication matrix. Then $\mathrm{rk}(M_i) = 1$ (why?) and $M_f = \sum_{i=1}^{\chi} M_i$. Thus, $\mathrm{rk}(M_f) \leq \sum_{i=1}^{\chi} \mathrm{rk}(M_i) \leq \chi \leq \chi(f)$. □

In order to fully appreciate how useful Theorem 4 is, let us note that M_{EQ_N} is the identity matrix (we tacitly assume that if $X = Y$ then the orders on rows and columns are consistent) and, therefore, $\mathrm{rk}(M_{\mathrm{EQ}_N}) = N$. This immediately gives Corollary 2. M_{LE_N} is the upper triangular matrix, and therefore we also have $\mathrm{rk}(M_{\mathrm{LE}_N}) = N$. Exercise 1 follows. It does require a little bit of thinking to see that the communication matrix M_{DISJ_n} is non-singular, that is $\mathrm{rk}(M_{\mathrm{DISJ}_n}) = 2^n$. But once it is done, we immediately obtain $C(\mathrm{DISJ}_n) \geq n$ which is essentially tight by (1) and also *stronger* than what we could do with combinatorial methods in Exercise 2 (the Ω is gone).

How tight is the bound of Theorem 4? It had been conjectured for a while that perhaps $\chi(f) \leq (\mathrm{rk}(M_f))^{O(1)}$ or maybe even $\chi(f) \leq O(\mathrm{rk}(M_f))$. In this form the conjecture was disproved in the series of papers [2, 23, 21]. But it is still possible and plausible that, say,

$$\chi(f) \leq 2^{O(\log_2 \mathrm{rk}(M_f))^2} ;$$

note that in combination with Theorem 3 that would still give a highly non-trivial inequality $C(f) \leq O(\log_2 \mathrm{rk}(M_f))^4$.

Despite decades of research, we still do not know the answer, and we actually do not have a very good clue how to even approach this problem that has become notoriously known as the *Log-Rank Conjecture*:

Open Problem 2 (Log-Rank Conjecture). *Is it true that*

$$\chi(f) \leq 2^{(\log_2 \mathrm{rk}(M_f))^{O(1)}} ?$$

Equivalently (by Theorems 1, 3), is it true that $C(f) \leq (\log_2 \mathrm{rk}(M_f))^{O(1)}$?

5 Probabilistic Models

This is all we wanted to say about the basic model of communication complexity. Even more fascinating and difficult problems arise when we introduce some variations. The most important of them, and the only one that we treat in sufficient detail in the rest of this article, is the model of *probabilistic communication complexity*.

Assume that Alice and Bob are now slightly less ambitious and agree to tolerate some small probability of error when computing the value of $f(x, y) \in \{0, 1\}$. Both of them are equipped with a fair unbiased coin (scientifically known as *generator of random bits*) that they can toss during the execution of the protocol, and adjust the messages they send to each other according to the result. Everything else is the same (that is, as on Figure 1) but we have to specify what it means that the protocol P correctly computes the function f.

Fix an input (x, y) and assume that Alice and Bob together flip their coins r times during the execution, which gives 2^r possible outcomes of these coin tosses. Some of them are *good* in the sense that Bob outputs the correct value $f(x, y)$, but some are *bad* and he errs. Let $\text{Good}(x, y)$ be the set of all good outcomes, then the quantity

$$p_{xy} = \frac{|\text{Good}(x, y)|}{2^r} \tag{4}$$

is for obvious reasons called the *probability of success* on the input (x, y).

What do we want from it? There is a very simple protocol of cost 1 that achieves $p_{xy} = 1/2$: Bob simply tosses his coin and claims that its outcome is $f(x, y)$. Thus, we definitely want to demand that

$$p_{xy} > 1/2 \,. \tag{5}$$

But how well should the probability of success be separated from $1/2$?

It turns out that there are essentially only three different possibilities (remember that we are lazy and do not care much about exact values of our constants). In the most popular and important version we require that $p_{xy} \geq 2/3$ for *any* input (x, y). The minimal cost of a probabilistic protocol that meets this requirement is called *bounded-error probabilistic communication complexity of the function f* and denoted by $R(f)$. If for any input pair (x, y) we only require (5) then the model is called *unbounded-error*, and the corresponding complexity measure is denoted by $U(f)$. In the third model (that is less known and will not be considered in our article), we still require (5), but now Alice and Bob are also charged for coin tosses. This e.g. implies that in any protocol of cost $O(\log_2 n)$, (5) automatically implies the better bound $p_{x,y} \geq \frac{1}{2} + \frac{1}{p(n)}$ for some polynomial $p(n)$.

Why, in the definition of $R(f)$, did we request that $p_{xy} \geq 2/3$, not $p_{xy} \geq 0.9999$? By using quite a general technique called *amplification*, it can be shown not to be very important. Namely, assume that Alice and Bob have at their disposal a protocol of cost $R(f)$ that achieves $p_{xy} \geq 2/3$, they repeat it independently 1000 times and output at the end the most frequent answer. Then the error probability of this repeated protocol of cost only $1000\,R(f)$ will not exceed 10^{-10}... (In order to prove this statement, some knowledge of elementary probability theory, like Chernoff bounds, is needed.)

Are coins really helpful for anything, that is are there any interesting problems that can be more efficiently solved using randomization than without it? An ultimate answer to this question is provided by the following beautiful construction, usually attributed to Rabin and Yao, that has to be compared with Corollary 2.

Theorem 5. $R(\mathrm{EQ}_N) \leq O(\log_2 \log_2 N)$.

Proof. In order to prove this, it is convenient to represent elements of X and Y as binary strings of length n, where $n = \lceil \log_2 N \rceil$. Furthermore, we want to view the binary string $x_1 x_2 \ldots x_n$ as the polynomial $x_1 + x_2 \xi + \ldots + x_n \xi^{n-1}$ in a variable ξ. Thus, Alice and Bob hold two polynomials $g(\xi)$ and $h(\xi)$ of the form above, and they want to determine if these polynomials are equal. For doing that they agree beforehand on a fixed prime number $p \in [3n, 6n]$ (such a prime always exists by Chebyshev's famous theorem). Alice tosses her coin to pick a random element $\xi \in \{0, 1, \ldots, p-1\}$. Then she computes the remainder (!) $g(\xi) \mod p$ and sends the pair $(\xi,\ g(\xi) \mod p)$ to Bob. Bob evaluates $h(\xi) \mod p$ and outputs 1 if and only if $h(\xi) \mod p$ is equal to the value $g(\xi) \mod p$ he received from Alice.

The cost of this protocol is only $O(\log_2 n)$ as required: this is how many *bits* you need to transmit a pair of integers $(\xi,\ g(\xi) \mod p)$ not exceeding $p \leq O(n)$ each. What about the probability of success? If $\mathrm{EQ}(g, h) = 1$ then $g = h$ and Bob clearly always outputs 1, there is no error in this case at all. But what will happen if $g \neq h$? Then $(h - g)$ is a *non-zero* polynomial of degree at most n. And any such polynomial can have at most n different roots in the finite field $\mathbb{F}_p$. If you do not understand the last sentence, then you can simply trust me that the number of bad $\xi \in \{0, 1, \ldots, p-1\}$, i.e., those for which Bob is fooled by the fact $g(\xi) = h(\xi) \mod p$, does not exceed $n \leq \frac{p}{3}$. And since ξ was chosen completely at random from $\{0, 1, \ldots, p-1\}$, this precisely means that the probability of success is at least $2/3$. $\square$

Let us now review other problems that we already saw before in the light of probabilistic protocols. The function less-or-equal from Exercise 1 also gives in to such protocols: $R(\mathrm{LE}_N) \leq O(\log_2 \log_2 N)$, although the proof is *way* more complicated than for equality [17, Exercise 3.18]. On the other hand, randomization does not help much for computing the disjointness function [4, 12, 24]:

Theorem 6. $R(\mathrm{DISJ}_n) \geq \Omega(n)$.

The proof is too complicated to discuss here. It becomes slightly easier for another important function, *inner product mod 2*, that we now describe.

Given $x, y \in \{0,1\}^n$, we consider, like in the case of disjointness, the set of all indices i for which $x_i = 1$ and $y_i = 1$. Then $\text{IP}_n(x, y) = 1$ if the cardinality of this set is odd, and $\text{IP}_n(x, y) = 0$ if it is even. Chor and Goldreich [9] proved the following:

Theorem 7. $R(\text{IP}_n) \geq \Omega(n)$.

The full proof is still too difficult to be included here, but we would like to highlight its main idea.

So far we have been interested only in f-monochromatic rectangles, i.e., those that are composed either of zeros only or of ones only. We typically wanted to prove that every such rectangle is small in a sense. In order to tackle probabilistic protocols, we need to consider *arbitrary* rectangles R. Every such rectangle has a certain number $N_0(f, R)$ of points with $f(x, y) = 0$, and $N_1(f, R)$ points with $f(x, y) = 1$. We need to prove that even if R is "large" then it is still "well balanced" in the sense that $N_0(f, R)$ and $N_1(f, R)$ are "close" to each other. Mathematically, the *discrepancy under uniform distribution*[5] of the function $f : X \times Y \longrightarrow \{0, 1\}$ is defined as

$$\text{Disc}_u(f) = \max_R \frac{|N_0(f, R) - N_1(f, R)|}{|X| \times |Y|},$$

where the maximum is taken over all possible combinatorial rectangles $R \subseteq X \times Y$.

It turns out that

$$R(f) \geq \Omega(\log_2(1/\text{Disc}_u(f))), \qquad (6)$$

that is, low discrepancy implies good lower bounds for probability protocols. Then the proof of Theorem 7 is finished by proving $\text{Disc}_u(\text{IP}_n) \leq 2^{-n/2}$ (which is rather non-trivial).

What happens if we go further and allow probabilistic protocols with unbounded error, that is we only require the success probability (4) to be strictly greater than $1/2$? The complexity of the equality function deteriorates completely [20]:

Theorem 8. $U(\text{EQ}_N) \leq 2$.

The disjointness function also becomes easy, and this is a good exercise:

Exercise 3. *Prove that* $U(\text{DISJ}_n) \leq O(\log_2 n)$.

The inner product, however, still holds the fort:

Theorem 9. $U(\text{IP}_n) \geq \Omega(n)$.

This result by Forster [10] is extremely beautiful and ingenious, and it is one of my favorites in the whole complexity theory.

---

[5] This concept can be generalized to other distributions as well.

6 Other Variations

We conclude with briefly mentioning a few modern directions in communication complexity where current research is particularly active.

6.1 Quantum Communication Complexity

Well, I will not even attempt to define what *quantum computers* are or if they have anything to do with the Quantum of Solace — most readers have probably heard of these still imaginary devices. Let me just say that they can be utilized for solving communication problems as well [28] and denote by $Q(f)$ the corresponding complexity measure. Quantum computers have an implicit access to random bits that implies $Q(f) \leq R(f)$. On the other hand, the discrepancy lower bound (6) still holds for quantum protocols [16] that gives for them the same bound as in Theorem 7. Something more interesting happens to the disjointness function: its complexity drops from n to $\sqrt{n}$ [7, 25]. Can a quantum communication protocol save more than a quadratic term over the best probabilistic protocol? This is one of the most important and presumably very difficult problems in the area:

Open Problem 3. *Is it true that $R(f)$ is bounded by a polynomial in $Q(f)$ for functions $f : X \times Y \longrightarrow \{0,1\}$?*

6.2 Multiparty Communication Complexity

Now we have more than 2 players, Alice, Bob, Claire, Dylan, Eve..., who collectively want to evaluate some function f. Depending on how the input to f is distributed among the players, there are several different models, the simplest being the scenario in which every player is holding her own set of data not known by any of the others. It turns out, however, that the most important one of them (by the token of having a *really* great deal of various applications) is the following *number-on-the-forehead* model. In this model, k players still want to evaluate a function $f(x^1, \ldots, x^k)$, $x^i \in \{0,1\}^n$. An interesting twist is that the i-th player has x^i written on his forehead, so he can actually see *all pieces of the input except for his own*. Let $C^k(f)$ as always be the minimal number of bits the players have to exchange to correctly compute $f(x^1, \ldots, x^k)$; for simplicity we assume that every message is broadcast to all other players at once.

Our basic functions DISJ_n and IP_n have "unique" natural generalizations DISJ_n^k and IP_n^k in this model. (Can you fill in the details?) The classical paper [5] proved the following bound:

Theorem 10. $C^k(\mathrm{IP}_n^k) \geq \Omega(n)$ *as long as $k \leq \varepsilon \log_2 n$ for a sufficiently small constant $\varepsilon > 0$.*

If we only could improve this result to a larger number of players (even for any other "good" function f), that would give absolutely fantastic consequences in complexity theory, some of which are outlined already in [5]. But this seems to be well out of reach of all methods that we currently have at our disposal.

Open Problem 4. *Prove that $C^k(\text{IP}_n^k) \geq n^\varepsilon$ for, say, $k = \lceil (\log_2 n)^2 \rceil$ and some fixed constant $\varepsilon > 0$.*

The multiparty communication complexity of DISJ_n^k was completely unknown for quite a while even for $k = 3$. A very recent breakthrough [8, 18, 6] gives lower bounds on $C^k(\text{DISJ}_n^k)$ that are non-trivial up to $k = \varepsilon(\log_2 n)^{1/3}$ players.

6.3 Communication Complexity of Search Problems

So far we have been considering functions that assume only two values, 0 and 1. In complexity theory such functions are often identified with *decision problems* or *languages*. But we can also consider functions of more general form $f : X \times Y \longrightarrow Z$, where Z is some more complicated finite set. Or we can go even one step further and assume that the function f is *multi-valued*, or in other words, we have a ternary relation $R \subseteq X \times Y \times Z$ such that for any pair (x, y) there exists at least one $z \in Z$ (a "value" of the multi-valued function f) such that $(x, y, z) \in R$. Given (x, y), the protocol P is supposed to output *some* $z \in Z$ with the property $(x, y, z) \in R$. Otherwise this z can be arbitrary. This kind of problems is called *search problems*.

The complexity of search problems is typically even more difficult to analyze than the complexity of decision problems. Let us consider just one important example, somewhat inspired by the equality function.

Assume that $X, Y \subseteq \{0, 1\}^n$, but that these sets of strings are disjoint: $X \cap Y = \emptyset$. Then $\text{EQ}(x, y) = 0$ for any $x \in X$, $y \in Y$ and there always exists a position i where they differ: $x_i \neq y_i$. Assume that the task of Alice and Bob is to actually *find* any such position.

This innocently-looking communication problem turns out to be equivalent to the second major open problem in computational complexity concerning computational depth [15, 22] (the first place being taken by the famous **P** vs. **NP** question). We do not have any clue as to how to prove lower bounds here. A simpler problem is obtained in a similar fashion from the disjointness function. That is, instead of $X \cap Y = \emptyset$ we assume that for any input $(x, y) \in X \times Y$ there is a position i such that $x_i = y_i = 1$. The task of Alice and Bob is once again to exhibit any such i. Lower bounds for this problem were indeed proved in [15, 22, 14], and they lead to very interesting consequences about the *monotone* circuit depth of Boolean functions.

7 Conclusion

In this article we tried to give some impression of how soon simple, elementary and innocent questions turn into open problems that have been challenging us for decades. There are even more such challenges in the field of computational complexity, and we are in the need of young and creative minds to answer these challenges. If this article has encouraged at least some of the readers to look more closely into this fascinating subject, the author considers its purpose fulfilled in its entirety.

List of Notation

Since this text uses quite a bit of notation, some of the most important notations are collected here together with a brief description, as well as the page of first appearance.

Complexity Measures

$\text{cost}(P)$ *cost of protocol P* — maximal number of bits to transmit in order to calculate the value of a function on any input (x, y) using protocol P **101**

$C(f)$ *(worst-case) communication complexity of function f* — minimal cost of any protocol computing f **101**

$\chi(f)$ *partition number of function f* — minimal number of pairwise disjoint f-monochromatic rectangles covering domain of f **105**

$t(f)$ *cover number of function f* — minimal number of f-monochromatic rectangles covering domain of f **107**

$\chi_0(f)$ minimal number of pairwise disjoint f-monochromatic rectangles covering $f^{-1}(\{0\})$ **107**

$\chi_1(f)$ minimal number of pairwise disjoint f-monochromatic rectangles covering $f^{-1}(\{1\})$ **107**

$t_0(f)$ minimal number of f-monochromatic rectangles covering $f^{-1}(\{0\})$ **107**

$t_1(f)$ minimal number of f-monochromatic rectangles covering $f^{-1}(\{1\})$ ($\log_2 t_1(f)$ *is called non-deterministic communication complexity of f*)   **107**

$R(f)$ *bounded-error probabilistic communication complexity of function f* — minimal cost of randomized protocol that assures that for any input the output will be correct with probability at least $\frac{2}{3}$ **109**

$U(f)$ *unbounded-error probabilistic communication complexity of function f* — minimal cost of randomized protocol that assures that for any input the output will be correct with probability greater than $\frac{1}{2}$ **109**

$\text{Disc}_u(f)$ *discrepancy (under uniform distribution) of function f* — maximal difference of how often values 0 and 1 occur on any rectangle (divided by $|X \times Y|$, where $X \times Y$ is the domain of f)   **111**

$Q(f)$ *quantum communication complexity of function f* — minimal cost of quantum computer protocol evaluating f **112**

$C^k(f)$ *multi-party communication complexity of function f* — minimal number of bits that k players have to transmit in order to correctly compute the value of f (in number-on-the-forehead model)   **112**

Binary Functions

EQ_N *equality function* — maps $\{1, 2, \ldots, N\} \times \{1, 2, \ldots, N\}$ to $\{0, 1\}$ with $\text{EQ}_N(x, y) = 1$ iff $x = y$ **102**

LE_N *less-or-equal function* — maps $\{1, 2, \ldots, N\} \times \{1, 2, \ldots, N\}$ to $\{0, 1\}$ with $\text{LE}_N(x, y) = 1$ iff $x \leq y$ **105**

DISJ_n *disjointness function ("NAND")* — maps $\{0, 1\}^n \times \{0, 1\}^n$ to $\{0, 1\}$ with $\text{DISJ}_n(x, y) = 1$ iff for all $i \leq n$ we have $x_i = 0$ or $y_i = 0$ **106**

IP_n *inner product mod 2* — maps $\{0, 1\}^n \times \{0, 1\}^n$ to $\{0, 1\}$ with $\text{IP}_n(x, y) = 1$ iff $x_i = y_i = 1$ for an odd number of indices i **111**

DISJ_n^k *generalized disjointness function* — maps $(\{0, 1\}^n)^k$ to $\{0, 1\}$ with $\text{DISJ}_n^k(x^1, \ldots, x^k) = 1$ iff for all $i \leq n$ there exists $\nu \in \{1, \ldots, k\}$ with $x_i^\nu = 0$ **112**

IP_n^k *generalized inner product mod 2* — maps $(\{0, 1\}^n)^k$ to $\{0, 1\}$ with $\text{IP}_n^k(x^1, \ldots, x^k) = 1$ iff the number of indices $i \leq n$ for which $x_i^1 = x_i^2 = \ldots = x_i^k = 1$ is odd   **112**

Growth of Functions[6] and Other

$O(f(n))$ $g(n) \leq O(f(n))$ iff there is $C > 0$ with $g(n) \leq Cf(n)$ for all n **102**

$\Omega(f(n))$ $g(n) \geq \Omega(f(n))$ iff there is $\varepsilon > 0$ with $g(n) \geq \varepsilon f(n)$ for all n **106**

$\lceil x \rceil$ the smallest integer $n \geq x$, for $x \in \mathbb{R}$ **101**

---

[6] The more traditional notation is $g(n) = O(f(n))$ and $g(n) = \Omega(f(n))$; see also footnote 3.

References

[1] Alfred V. Aho, Jeffrey D. Ullman, and Mihalis Yannakakis, On notions of information transfer in VLSI circuits. In: *Proceedings of the 15th ACM Symposium on the Theory of Computing*, pp. 133–139. ACM Press, New York (1983)

[2] Noga Alon and Paul Seymour, A counterexample to the rank-coloring conjecture. *Journal of Graph Theory* **13**, 523–525 (1989)

[3] Sanjeev Arora and Boaz Barak, *Computational Complexity: A Modern Approach*. Cambridge University Press, Cambridge (2009)

[4] László Babai, Peter Frankl, and Janos Simon, Complexity classes in communication complexity theory. In: *Proceedings of the 27th IEEE Symposium on Foundations of Computer Science*, pp. 337–347. IEEE Computer Society, Los Alamitos (1986)

[5] László Babai, Noam Nisan, and Márió Szegedy, Multiparty protocols, pseudorandom generators for logspace, and time-space trade-offs. *Journal of Computer and System Sciences* **45**, 204–232 (1992)

[6] Paul Beame and Dang-Trinh Huynh-Ngoc, Multiparty communication complexity and threshold circuit size of AC^0. Technical Report TR08-082, *Electronic Colloquium on Computational Complexity* (2008)

[7] Harry Buhrman, Richard Cleve, and Avi Wigderson, Quantum vs. classical communication and computation. In: *Proceedings of the 30th ACM Symposium on the Theory of Computing*, pp. 63–86. ACM Press, New York (1998). Preliminary version available at http://arxiv.org/abs/quant-ph/9802040

[8] Arkadev Chattopadhyay and Anil Ada, Multiparty communication complexity of disjointness. Technical Report TR08-002, *Electronic Colloquium on Computational Complexity* (2008)

[9] Benny Chor and Oded Goldreich, Unbiased bits from sources of weak randomness and probabilistic communication complexity. *SIAM Journal on Computing* **17**(2), 230–261 (1988)

[10] Jürgen Forster, A linear lower bound on the unbounded error probabilistic communication complexity. *Journal of Computer and System Sciences* **65**(4), 612–625 (2002)

[11] Martin Fürer, Faster integer multiplication. *SIAM Journal on Computing* **39**(3), 979–1005 (2009)

[12] Bala Kalyanasundaram and Georg Schnitger, The probabilistic communication complexity of set intersection. *SIAM Journal on Discrete Mathematics* **5**(4), 545–557 (1992)

[13] Anatolii A. Karatsuba and Yuri P. Ofman, Multiplication of many-digital numbers by automatic computers. *Proceedings of the USSR Academy of Sciences* **145**, 293–294 (1962)

[14] Mauricio Karchmer, Ran Raz, and Avi Wigderson, Super-logarithmic depth lower bounds via direct sum in communication complexity. *Computational Complexity* **5**, 191–204 (1995)

[15] Mauricio Karchmer and Avi Wigderson, Monotone circuits for connectivity require super-logarithmic depth. *SIAM Journal on Discrete Mathematics* **3**(2), 255–265 (1990)

[16] Ilan Kremer, *Quantum Communication*. Master's thesis, Hebrew University, Jerusalem (1995)

[17] Eyal Kushilevitz and Noam Nisan, *Communication Complexity*. Cambridge University Press, Cambridge (1997)

[18] Troy Lee and Adi Shraibman, Disjointness is hard in the multiparty number-on-the-forehead model. *Computational Complexity* **18**(2), 309–336 (2009)

[19] Kurt Mehlhorn and Erik M. Schmidt, Las Vegas is better than determinism in VLSI and distributive computing. In: *Proceedings of the 14th ACM Symposium on the Theory of Computing*, pp. 330–337. ACM Press, New York (1982)
[20] Ramamohan Paturi and Janos Simon, Probabilistic communication complexity. *Journal of Computer and System Sciences* **33**(1), 106–123 (1986)
[21] Ran Raz and Boris Spieker, On the "log-rank"-conjecture in communication complexity. *Combinatorica* **15**(4), 567–588 (1995)
[22] Alexander Razborov, Applications of matrix methods to the theory of lower bounds in computational complexity. *Combinatorica* **10**(1), 81–93 (1990)
[23] Alexander Razborov, The gap between the chromatic number of a graph and the rank of its adjacency matrix is superlinear. *Discrete Mathematics* **108**, 393–396 (1992)
[24] Alexander Razborov, On the distributional complexity of disjointness. *Theoretical Computer Science* **106**, 385–390 (1992)
[25] Alexander Razborov, Quantum communication complexity of symmetric predicates. *Izvestiya: Mathematics* **67**(1), 145–159 (2003)
[26] Arnold Schönhage and Volker Strassen, Schnelle Multiplikation großer Zahlen. *Computing* **7**, 281–292 (1971)
[27] Andrew Yao, Some complexity questions related to distributive computing. In: *Proceedings of the 11th ACM Symposium on the Theory of Computing*, pp. 209–213. ACM Press, New York (1979)
[28] Andrew Yao, Quantum circuit complexity. In: *Proceedings of the 34th IEEE Symposium on Foundations of Computer Science*, pp. 352–361. IEEE Computer Society, Los Alamitos (1993)

Ten Digit Problems

Lloyd N. Trefethen

Abstract. Most quantitative mathematical problems cannot be solved exactly, but there are powerful algorithms for solving many of them numerically to a specified degree of precision like ten digits or ten thousand. In this article three difficult problems of this kind are presented, and the story is told of the SIAM 100-Dollar, 100-Digit Challenge. The twists and turns along the way illustrate some of the flavor of algorithmic continuous mathematics.

1 Introduction

I am a mathematician who spends his time working with numbers, real numbers like $0.3233674316\ldots$ and $22.11316746\ldots$. If I can compute a quantity to ten digits of accuracy, I am happy. Most mathematicians are not like this! In fact, sometimes it seems that the further you go in mathematics, the less important actual numbers become. But some of us develop algorithms to solve problems quantitatively, and we are called numerical analysts. I am the head of the Numerical Analysis Group at Oxford.

Like all mathematicians, I enjoy having a concrete problem to chew on. For example, what is the value of the integral

$$\int_0^1 x^{-1} \cos(x^{-1} \log x) \, dx \,? \tag{1}$$

You won't find the answer in a table of integrals, and I don't think anybody knows how to derive an exact formula. But even though an exact formula does not exist, the integral still makes sense. (More precisely, it makes sense

---

Lloyd N. Trefethen
Oxford University Mathematical Institute, 24–29 St Giles, Oxford OX1 3LB, UK.
e-mail: trefethen@maths.ox.ac.uk

if we define (1) as the limit $\varepsilon \to 0$ of an integral from ε to 1.) The only way to evaluate it is by some kind of numerical algorithm, and this is hard, for the integrand (i.e., the function under the integral sign) oscillates infinitely often as x approaches 0 while swinging between larger and larger values that diverge to infinity. To ten digits, the answer turns out to be the first number listed above.

Each October in Oxford, four or five new graduate students arrive to begin a PhD in numerical analysis, and in their first term they participate in a course called the Problem Solving Squad. Each week I give them a problem like (1) whose solution is a single real number. Working in pairs, their job is to compute this number to as many digits of accuracy as they can. I don't give any hints, but the students are free to talk to anybody and use the library and the web. By the end of six weeks we always have some unexpected discoveries — and some tightly bonded graduate students!

In this article I want to tell you about three of these problems that have given me pleasure, which I'll call "two cubes", "five coins", and "blowup". Though this is the first article I've written about it, Oxford's Numerical Analysis Problem Solving Squad has been well known since the *SIAM 100-Dollar, 100-Digit Challenge* was organized in 2002. This involved ten problems selected from the early years of the Squad, and the challenge for contestants was to try to solve each problem to as many digits of accuracy as possible, up to ten digits for each. Teams from around the world entered the race, and twenty of them achieved perfect scores of 100 and won $100. Afterwards a book about the problems was published by four of the winners, with a cover picture illustrating an ingenious method of solving (1) using complex numbers [1]. I'll say more about the 100-Digit Challenge at the end.

2 Two Cubes

Our first problem is motivated by a simple question from physics. Isaac Newton discovered that if two point masses of magnitude m_1 and m_2 are separated by a distance r, then they are attracted towards each other by a gravitational force of magnitude
$$F = \frac{Gm_1 m_2}{r^2},$$
where G is a constant known as the *gravitational constant*. If you have masses that are not points but spheres or other objects, then each point in one mass is attracted to each point in the other by the same formula. We now pose the following idealized problem:

Problem 1. *Two objects of mass 1 attract each other gravitationally according to Newton's law with $G = 1$. Each object is a unit cube with its mass uniformly distributed. The centers of the cubes are one unit apart, so they are in contact along one face. What is the total force, F?*

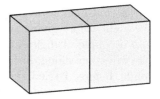

You can think of the cubes as suns or planets if you like. No mathematician will be troubled by the idea of cubic planets. That's what mathematics is best at, reasoning about any kind of precisely defined situation, no matter how artificial. Something else about this problem, however, marks it out as unusual for most mathematicians. *It is so trivial!* We all know the formula for gravity, so where's the interest here? Working out the force between these particular bodies should be just a matter of bookkeeping. We are not in this business to be bookkeepers!

But some of us are in this business to design algorithms, and this innocent-looking problem is a killer. Let's try to solve it, and you'll see what I mean.

The first thought that may occur to you is, can't we replace each cube by a unit mass at the center and get the answer $F = 1$? Isn't that what Newton showed so many years ago, that as far as gravity is concerned, planets are equivalent to points? Well yes, Newton did show that, but only for spherical planets. If the shape is a cube, we have to investigate more carefully.

Let's say that cube 1 consists of points (x_1, y_1, z_1) with $0 < x_1, y_1, z_1 < 1$, and cube 2 consists of points (x_2, y_2, z_2) in the same range except $1 < x_2 < 2$. For unit point masses at (x_1, y_1, z_1) and (x_2, y_2, z_2), the force would be

$$\frac{1}{r^2} = \frac{1}{(x_1 - x_2)^2 + (y_1 - y_2)^2 + (z_1 - z_2)^2},$$

aligned in the direction between the points. Our job is to add up these forces over all pairs of points (x_1, y_1, z_1) and (x_2, y_2, z_2). That is, we need to evaluate a *six-dimensional integral*. The y and z components of the total force will cancel to zero, by symmetry, so it's the x component we need to integrate, which is equal to $(x_2 - x_1)/r$ times the expression above. That is, the x component of the force between unit masses at (x_1, y_1, z_1) and (x_2, y_2, z_2) is

$$f(x_1, y_1, z_1, x_2, y_2, z_2) = \frac{x_2 - x_1}{[(x_1 - x_2)^2 + (y_1 - y_2)^2 + (z_1 - z_2)^2]^{3/2}}. \quad (2)$$

The number F we are looking for is thus

$$F = \int_0^1 \int_0^1 \int_1^2 \int_0^1 \int_0^1 \int_0^1 f(x_1, \ldots, z_2) \, dx_1 \, dy_1 \, dz_1 \, dx_2 \, dy_2 \, dz_2. \quad (3)$$

This is an integral over a six-dimensional cube. How do we turn it into a number?

I must be honest and confess that I am always behind schedule and usually make up these problems the night before giving them to the students. I aim to make sure I can compute a couple of digits at least, trusting that more powerful and beautiful ideas will come along during the week.

For this problem, the night before, I tried the most classical numerical method for evaluating integrals, *Gauss quadrature*. The idea of Gauss quadrature in one dimension is to sample the integrand at n precisely defined values of x called *nodes*, multiply the sampled values by n corresponding real numbers called *weights*, and add up the results. (The nodes and weights are determined by the condition that the estimate comes out exactly correct if the integrand happens to be a polynomial of degree no greater than $2n - 1$.) For smooth integrands, such as those defined by functions that can be differentiated several times, this gives amazingly accurate approximations. And by squaring or cubing the grid, you can evaluate integrals in two or three dimensions. Here are pictures of 10, 10^2, and 10^3 Gauss nodes for integration over an interval, a square, and a cube:

For our integral (3) the same idea applies, though it's not so easy to draw a picture.

Here is what I found with this method of "Gauss quadrature raised to the 6-th power". The number of nodes is $N = n^6$ with $n = 5, 10, \ldots, 30$, FN is the Gauss quadrature approximation to F, and time is the amount of time each computation took on my computer.

```
N =       15625    FN = 0.969313    time =     0.0 secs.
N =      1000000   FN = 0.947035    time =     0.3 secs.
N =     11390625   FN = 0.938151    time =     3.2 secs.
N =     64000000   FN = 0.933963    time =    17.6 secs.
N =    244140625   FN = 0.931656    time =    66.7 secs.
N =    729000000   FN = 0.930243    time =   198.2 secs.
```

This is awful! We can see that the answer looks like $F \approx 0.93$, 7% less than if the cubes were spheres. But that is all we can see, and it has taken minutes of computing time. Computing 10 digits would take pretty much forever.

In fact, these results from Gauss quadrature with its special nodes and weights are worse than what you get if you set all the weights equal to $1/N$ and place the nodes at random in the six-dimensional cube! This kind of randomized computation is called the *Monte Carlo method*. Here are typical sets of 10, 100, and 1000 random nodes in one, two and three dimensions:

. .   ... .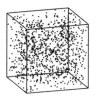

And here's a set of Monte Carlo results for the same values of N as before.

```
N =      15625    FN = 0.906741    time =   0.1 secs.
N =    1000000    FN = 0.927395    time =   0.5 secs.
N =   11390625    FN = 0.925669    time =   4.4 secs.
N =   64000000    FN = 0.925902    time =  22.7 secs.
N =  244140625    FN = 0.926048    time =  88.0 secs.
N =  729000000    FN = 0.925892    time = 257.0 secs.
```

It seems that we now have three or four digits, $F \approx 0.9259$ or 0.9260. In this collection of results, and indeed for all the numbers reported in this article, it is a very interesting matter to try to make more precise statements about the accuracy of a computation. This is an important aspect of the field of numerical analysis, but to keep things as simple as possible, we shall settle for experimental evidence here and not attempt such estimates.

So, the world's slickest method for numerical integration is losing out to the world's simplest! Actually this often happens with high-dimensional integrals. The errors with Monte Carlo decrease in proportion to $1/\sqrt{N}$, the inverse of the square root of the number of samples, more or less independently of the number of dimensions, whereas Gauss quadrature slows down greatly with increasing dimension. This is a widespread theme in numerical algorithms, and one speaks of "the curse of dimensionality".

But even Monte Carlo hits a wall at 4 or 5 digits, or maybe 6 or 7 if we run overnight or use a parallel computer. How can we get more? The students worked hard and came up with many good ideas. Let's focus on one of these which eventually turned into a ten digit solution.

If you're familiar with Gauss quadrature, you can quickly spot why it has done so badly. The problem is that the integrand (2) is not smooth but *singular* because the cubes are right up against each other. The denominator goes to zero whenever $x_1 = x_2 = 1$, $y_2 = y_1$, and $z_2 = z_1$, so the fraction goes to ∞. This isn't bad enough to make the values of the integral infinite, but it slows down the convergence terribly.

We would like to eliminate the singularity. One way to do it would be to change the problem by separating the cubes, say, by a distance 1.

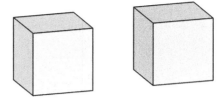

The convergence of Gauss quadrature changes completely, giving 14 digits in a fraction of a second:

```
N =     15625    F = 0.24792296453612    time =  0.0 secs.
N =   1000000    F = 0.24792296916638    time =  0.3 secs.
N =  11390625    F = 0.24792296916638    time =  3.2 secs.
N =  64000000    F = 0.24792296916638    time = 17.6 secs.
```

Notice that the answer is close to 1/4, which is what the force would be if the cubes were spheres with centers separated by distance 2.

So we can accurately solve a modified problem, with the cubes separated. What about the original problem? Let $F(\varepsilon)$ denote the force between cubes separated by a distance $\varepsilon \geq 0$. We want to know $F(0)$, but we can only evaluate $F(\varepsilon)$ accurately for values of ε that are not too small. A good idea is to perform some kind of *extrapolation* from $\varepsilon > 0$ to $\varepsilon = 0$. Extrapolation is a well-developed topic in numerical mathematics, and some of the important methods in this area are known as Richardson extrapolation and Aitken extrapolation. The students and I tried a number of strategies like these and got... well, we were disappointed. We got another digit or two.

And then along came a delightful additional idea from graduate student Alex Prideaux, which finally nailed the two cubes problem.

Prideaux's idea was, let's break each cube into eight pieces, eight sub-cubes of size 1/2. Now the number F will be the sum of 64 pairwise contributions.

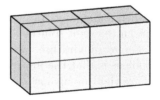

Four of these pairs meet along a face. Eight pairs meet along an edge, and four meet at a vertex:

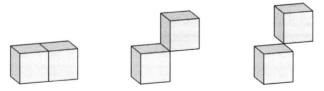

In the other 48 cases the sub-cubes are well separated.

Having started with one kind of six-dimensional quadrature problem, we now have four! — face, edge, vertex, and separated. Let's define Face(d) to be the x component of the force between two cubes of size d touching along a face, and similarly Edge(d) and Vertex(d) for cubes of size d touching along edges and at vertices. If you think about the picture with 16 sub-cubes above, you will see that we can write equations for the forces at scale 1 in terms of the forces at scale $\frac{1}{2}$ like this:

$$\text{Vertex}(1) = \text{Vertex}(\tfrac{1}{2}) + \text{well-separated terms},$$

$$\text{Edge}(1) = 2\,\text{Edge}(\tfrac{1}{2}) + 2\,\text{Vertex}(\tfrac{1}{2}) + \text{well-separated terms},$$

$$\text{Face}(1) = 4\,\text{Face}(\tfrac{1}{2}) + 8\,\text{Edge}(\tfrac{1}{2}) + 4\,\text{Vertex}(\tfrac{1}{2}) + \text{well-separated terms}.$$

This may look like dubious progress, until we note a basic fact of scaling:

$$\text{Vertex}(\tfrac{1}{2}) = \tfrac{1}{16}\text{Vertex}(1), \quad \text{Edge}(\tfrac{1}{2}) = \tfrac{1}{16}\text{Edge}(1), \quad \text{Face}(\tfrac{1}{2}) = \tfrac{1}{16}\text{Face}(1).$$

The factors of 16 come about as follows. If you halve the scale of a cubes problem, each mass decreases by a factor of 8, so the product of masses decreases by 64. On the other hand the distance between the cubes also cuts in half, so $1/r^2$ increases by a factor of 4. Thus overall, the force changes by the factor $4/64 = 1/16$.

Putting these observations together, we find

$$\text{Vertex}(1) = \tfrac{1}{16}\text{Vertex}(1) + \text{well-separated terms},$$

$$\text{Edge}(1) = \tfrac{2}{16}\text{Edge}(1) + \tfrac{2}{16}\text{Vertex}(1) + \text{well-separated terms},$$

$$\text{Face}(1) = \tfrac{4}{16}\text{Face}(1) + \tfrac{8}{16}\text{Edge}(1) + \tfrac{4}{16}\text{Vertex}(1) + \text{well-separated terms}.$$

We can calculate the well-separated terms to high accuracy in a second or two by Gauss quadrature, and these formulas give us first Vertex(1), then Edge(1), and then the number we care about, Face(1). The answer is

$$F \approx 0.9259812606\,.$$

3 Five Coins

The second problem involves no physics, just geometry and probability.

Problem 2. *Non-overlapping coins of radius 1 are placed at random in a circle of radius 3 until no more can fit. What is the probability p that there are 5 coins?*

We shall see that this story, so far at least, has a less happy ending.

We can illustrate the game in a picture. We put a red coin down at random, then a green one, then a blue one. I'll leave it to you to prove that fitting in at least three coins is always possible. Here's an example that gets stuck at three, with no room for a fourth.

(Incidentally, to be precise the problem must specify what "at random" means. The meaning is pretty obvious; the trick is how to say it mathematically. If k coins are down, consider the set S of points at which the center of another coin could be placed. If S is nonempty, we put the center of coin $k + 1$ at a point in S selected at random with respect to area measure.)

Quite often a fourth coin can fit too. Here's an example.

At four coins, we are usually finished. But occasionally a fifth can fit too:

Five coins is the limit. (Well, not quite. Six or seven are also possible, but the probabilities of these events are zero, meaning that no matter how many times you play the game randomly, you will never see them. Can you prove it? Think of where the centers of a 7-coin configuration would have to fall.)

So the question is, how often do we get to five coins? This problem has something in common with the two cubes. Since it is posed in terms of probability, one approach to a solution should be Monte Carlo simulation. We can write a computer program and see what happens. It's not obvious how best to organize the computation, but one reasonable approach is to replace the big disk by a fine grid, then pick points at random in that grid. Every time you pick a point, you remove it from further consideration along with all its neighbors at distance less than 2. For convergence to the required number, you must refine the grid and also increase the number of samples. By following this Monte Carlo approach we find approximately these frequencies:

3 coins: 18%   4 coins: 77%   5 coins: 5%

This level of accuracy can be achieved in 5 minutes. Running overnight gives possibly one more digit:

$$p \approx 0.053.$$

You may sense a bit of foreshadowing in the fact that we've printed this number in large type.

There's a scientific context to the five coins: this is known as a "parking problem". In one dimension, imagine a curb of length L with k cars parking at random, one after another, along the curb. How many cars will fit? Problems like this are of interest to chemists and physicists investigating aggregation of particles, and have been studied in 1, 2, and 3 dimensions. A question often asked is, in the limit of an infinite parking area, what fraction of the space can one expect will be filled by randomly arriving cars or coins or particles? In the one-dimensional case, the answer is known in the form of an integral that evaluates to 0.7475979202....

For circular disks ("coins") in two dimensions, or spheres in three, we speak of a "Tanemura parking problem". So far as I am aware, no formula is known for the infinite-size limit in either of these situations.

But in any case our Problem 2 concerns not a limit but a very concrete setting of 3, 4, and 5 coins. And do you know something? Despite hard work, the Problem Squad never improved on 0.053. We found variations on the theme of Monte Carlo, but none that helped decisively. Yet this problem is one of finite-dimensional geometry, equivalent in fact to another multiple integral. There must be a way to solve it to high accuracy!

Sometimes a problem has no slick solution. In this case, I think a slick solution is still waiting to be found.

4 Blowup

Our final problem involves a partial differential equation (PDE). Since this may be unfamiliar territory, let me explain a little.

One of the best known PDEs is the *heat* or *diffusion equation:*

$$\frac{\partial u}{\partial t} = \frac{\partial^2 u}{\partial x^2}. \tag{4}$$

We have here a function $u(x,t)$ of a space variable x and a time variable t. The equation says that at each point in space-time, the partial derivative of u with respect to t is equal to its second partial derivative with respect to x. Physically, the idea is that at a particular point x and time t, the temperature will increase ($\partial u/\partial t > 0$) if the temperature curves upward as a function of x ($\partial^2 u/\partial x^2 > 0$), since this means that heat will flow in towards x from nearby hotter points.

For example, one solution to (4) would be the function

$$u(x,t) = e^{-t}\sin(x),$$

for it is not hard to verify that for this choice of u,

$$\frac{\partial u}{\partial t} = -u, \qquad \frac{\partial^2 u}{\partial x^2} = -u.$$

In the time of Napoleon, Joseph Fourier discovered that equation (4) governs diffusion of heat in a one-dimensional body. For example, if a function $u_0(x)$ describes the temperature distribution in an infinitely long bar at time $t = 0$, then a solution $u(x,t)$ to (4) with initial condition $u(0,x) = u_0(x)$ tells the temperature at times $t > 0$. This was a first-class scientific discovery, and it is bad luck for Mr. Fourier that for whatever accidental reason of history, we talk about the Laplace and Poisson equations but not the Fourier equation.

Most PDE problems are posed on bounded domains, and then we must prescribe *boundary conditions* to determine the solution. For example, the heat equation might apply on the interval $x \in [-1,1]$, corresponding to a finite bar, with boundary conditions $u(-1,t) = u(1,t) = 0$, corresponding to zero temperature at both ends. Here's an illustration of a solution to this problem at different times. Notice how the sharp edges diffuse instantly, whereas the larger structure decays more slowly. That makes sense, since strong temperature differences between nearby points will quickly equalize.

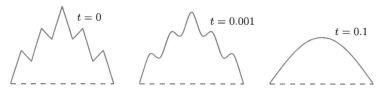

Eventually, all the heat flows out the ends and the signal decays to zero. (If you are troubled by how to take the second derivative of the jagged initial function, you are right to be troubled! We may imagine that $u(x,0)$ is a smooth function that happens to match the jagged curve to high accuracy.)

Equation (4) depends linearly on the variable u: it is a linear PDE. Our third problem involves a nonlinear PDE which consists of this equation plus an additional term, the exponential of u:

$$\frac{\partial u}{\partial t} = \frac{\partial^2 u}{\partial x^2} + e^u. \tag{5}$$

Whereas the heat equation just moves heat around, conserving the total heat apart from any inflow or outflow at the boundaries, this nonlinear term *adds* heat. You can think of e^u as a model of a chemical process like combustion, a temperature-dependent kind of combustion that accelerates exponentially as the temperature goes up.

Suppose we apply equation (5) on an interval $[-L, L]$ with initial condition $u(x, 0) = 0$ and boundary conditions $u(-L, t) = u(L, t) = 0$. For $t > 0$, the exponential term adds heat, and the derivative term diffuses it out the boundaries. There's a competition here. If L is small, diffusion wins and the solution converges to a fixed limit function $u_\infty(x)$ as $t \to \infty$, with combustion exactly balancing diffusion. If L is larger, combustion wins. The heat can't diffuse away fast enough, and the solution explodes to infinity at a finite time $t = t_c$. In particular, this happens for the case $L = 1$. Here are the solutions at times 0, 3, and 3.544, by which point the amplitude has reached about 7.5. Soon afterwards, the curve will explode to infinity.

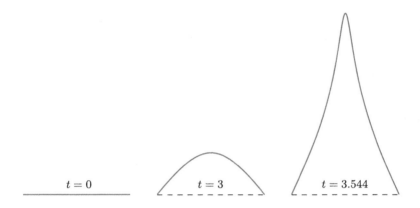

Physically, this blowup is related to the phenomenon of *spontaneous combustion*. Imagine a heap of grass cuttings or compost. Chemical processes may generate some heat, but if it's a small heap, the heat escapes and all is stable. In a larger pile, however, the heat may be unable to get away. Eventually, the heap catches fire. Much the same mathematics explains why a quantity of Uranium 235 has a *critical mass* above which it explodes in a fission chain reaction, the original principle behind atomic bombs.

Here then is our mathematical problem.

Problem 3. *At what time t_c does the solution $u(x, t)$ to the problem*

$$\frac{\partial u}{\partial t} = \frac{\partial^2 u}{\partial x^2} + e^u, \qquad u(x, 0) = 0, \quad u(-1, t) = u(1, t) = 0 \qquad (6)$$

blow up to infinity?

The numerical solution of PDEs on computers goes back to von Neumann and others in the 1940s. It is as important a topic as any in numerical analysis, and a huge amount is known. For Problem 3, the geometry is an interval and the equation is a simple one with a single variable. Other problems of interest to scientists and engineers may be much more complicated. The shapes of wings and airplanes are designed by solving PDEs of fluid and structural mechanics in complicated three-dimensional geometries. Weather forecasts come from solutions of PDE problems involving variables representing air velocity, temperature, pressure, humidity and more, and now the geometry is nothing less than a chunk of planet Earth with its oceans and islands and mountains.

Most numerical solutions of PDEs depend on discretizing the problem, replacing partial derivatives by finite approximations. The grid around an airplane may be eye-poppingly complicated, but for (6), one might begin by using a simple regular grid like this one, with the horizontal direction corresponding to x and the vertical direction to t.

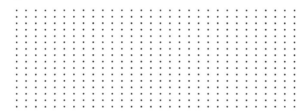

Quite a good solution strategy for Problem 3 is to discretize (6) on grids like this, shrink the step sizes Δx and Δt systematically, and then use some kind of extrapolation to estimate the blowup time.

For example, one way to discretize this equation from $t = 0$ to $t = 3.544$ is to divide $[-1, 1]$ into N space intervals and $[0, 3.544]$ into $2N^2$ time intervals and then approximate the PDE on this grid in a manner whose details we won't go into. Here are the approximate values $u(0, 3.544)$ produced by this method for a succession of values of N:

```
N =    32    u(0,3.544) = 9.1015726    time =    0.0 secs.
N =    64    u(0,3.544) = 7.8233770    time =    0.1 secs.
N =   128    u(0,3.544) = 7.5487013    time =    0.6 secs.
N =   256    u(0,3.544) = 7.4823971    time =    3.3 secs.
N =   512    u(0,3.544) = 7.4659568    time =   21.2 secs.
N =  1024    u(0,3.544) = 7.4618549    time =  136.2 secs.
```

It seems clear that the true value $u(0, 3.544)$ must be about 7.46, and by applying Richardson extrapolation to the data, one can improve this estimate to 7.460488. Using methods like this, with some ingenuity and care, one can estimate the blow-up time for Problem 3 to six or seven digits.

It seems wasteful, however, to use a regular grid for a problem like this where all the action happens in a narrow spike near $x = 0$ and $t = 3.5$. It is tempting to try to take advantage of this structure by using some kind of uneven grid, one which gets finer as the spike gets narrower, like this:

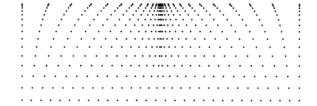

Adapting grids like this to the solution being computed is a big topic in numerical PDE. When flows are computed over airplanes, the grids may be thousands of times finer near the surface than further out.

The ten digit solution that I know of for Problem 3 makes use of a highly nontrivial adaptive-gridding algorithm due to my former student Wynn Tee. Tee's method starts from the observation that although (6) is posed for x in the interval $[-1, 1]$, we can extend the solution to complex values of x too, that is, to values of x with an imaginary as well as a real part. As t approaches t_c, it turns out that the solution $u(x, t)$ has singularities in the complex x-plane that approach the real axis. By monitoring this situation and using what is known as a conformal map to distort the grid systematically, one can maintain extremely high accuracy with a small number of grid points even as the spike grows very tall and narrow. In fact it is possible to calculate a solution to ten digits of accuracy with only 100 grid points in the x direction:

$$t_c \approx 3.544664598.$$

This solution also takes advantage of advanced methods of time discretization, and it is really a tour de force of clever computation, illustrating that some very abstract mathematics may be useful for concrete problems.

5 The 100-Digit Challenge

What does it mean to solve a mathematical problem? That's too big a question, for the solution to a mathematical problem might be a "yes" or "no", a proof, a counterexample, a theorem, who knows what. More specifically, then: what does it mean to solve one of those mathematical problems whose solution, in principle, is a number? Must we find an exact formula — and if we do, is that good enough regardless of the formula's complexity? Must we write down the number in decimal form — and how many digits are enough? Is it sufficient to find an algorithm that can generate the number — and does it matter how quickly it runs?

There are plenty of sand traps in this discussion. Even the notion of an exact formula is elusive. For example, in the theory of roots of polynomials, an exact formula is traditionally allowed to include n-th roots, like $\sqrt[3]{2}$, but not trigonometric or other special functions, like $\sin(2)$. A computer, however, doesn't see much difference between $\sqrt[3]{2}$ and sin(2). Both are calculated by fast iterative algorithms. For that matter, even a fraction like 2/3 is calculated by an iterative algorithm on some computers, but nobody would question the credentials of 2/3 as an exact solution! What about a more complicated expression like $\int_0^1 e^{-x^4} dx$? That's easy for your computer too, but it will probably have to call a piece of software rather than use something built-in to the microprocessor hardware. Is it an exact solution?

For me, solving a numerical problem means finding an algorithm that can calculate the answer to high accuracy on a computer, whether or not there's an explicit formula, and this brings us to the question: what's so special about ten digits? Why not three digits, or a hundred? I'd like to suggest two reasons why ten digits is indeed a good goal to aim for.

One reason is that in science, many things are known to more than three digits of accuracy, but hardly anything is known to more than ten. If a quantity is known to a hundred or a million digits, you can be sure it is a mathematical abstraction like π rather than a physical constant like the speed of light or Planck's constant. So in science, you might say that ten digits is more or less the same as infinity. Ultimately this is why computers normally compute with 16-digit precision, not 160. (And since 10 digits is comfortably less than 16, you usually don't have to worry too much about computer precision when tackling a ten-digit problem.)

The second reason is illustrated by the five coins. To exaggerate a little bit, you can solve just about any problem to three digits of accuracy by brute force. But a brute force algorithm doesn't encode much insight, and it often fails if you try to push it much further. This is just what happened with the five coins, where we got stuck at three digits with Monte Carlo. Ten digits is a very different achievement. To get to ten digits, you need a good understanding of your problem and a well targeted algorithm. In fact, if you can get this far, the chances are pretty good that you could get ten thousand if you had to. To see what I mean, let's return to the 100-Digit Challenge.

The Challenge was launched in January 2002, and its ten problems involved the integral (1), some chaotic dynamics, the norm of an infinite matrix, global optimization in two dimensions, the approximation of the gamma function in the complex plane, a random walk on a lattice, inverting a $20\,000 \times 20\,000$ matrix, the heat equation on a square plate, parametrized optimization, and Brownian motion. Each team was allowed to have up to six members, and 94 teams entered from 25 countries. Twenty of them got perfect scores! That surprised me. I had planned to spend \$100 rewarding whoever got the most digits, but with twenty perfect scores, I was unsure what to do. To my great pleasure, a donor came forward to plug the gap — William Browning, founder of Applied Mathematics, Inc. in Connecticut. You might think that for a member of a team of six to give nights and weekends to a mathematical project and be rewarded with \$16.67 must be some kind of sour joke. But it turned out that receiving a little bit of cash meant a great deal to the winners as a symbolic recognition of their achievement.

The winners also got certificates, like this one awarded to Folkmar Bornemann of the Technical University of Munich, one of the authors of the book *The SIAM 100-Digit Challenge* [1].

I published an article in *SIAM News* telling the story and outlining a solution to each problem. The article ended like this:

> If you add together our heroic numbers, the result is $\tau = 1.497258836\ldots$. I wonder if anyone will ever compute the ten thousandth digit of this fundamental constant?

Now I wrote this to be funny, and to get people thinking. What's funny is that this number τ is the most unfundamental constant you could imagine. The sum of the answers to ten arbitrary unrelated problems — what nonsense! I chose the Greek letter tau because privately I was thinking of this as "Trefethen's Constant". With good British modesty I felt it was OK to name something after oneself, provided the item was sufficiently ridiculous.

In the book [1], τ took on a life of its own. The authors amazed us all by finding ways to solve nine of the problems, one after another, to ten thousand digits! The variety of mathematical, algorithmic, and computational tools they used was striking. Indeed, the book emphasizes throughout that there is no "right way" to solve a problem, and your bag of tools can never be too big. By one beautiful chain of reasoning making use of ideas of the Indian mathematician Ramanujan, Bornemann found an exact formula for the solution to Challenge Problem 10 (Brownian motion). By another remarkable method based on ideas related to the field of number theory, Jean-Guillaume Dumas together with 186 computer processors running for four days were able to compute exactly the solution to Challenge Problem 7 (inverting a matrix): the task was to find a particular element of this inverse, and they discovered that the answer was a rational number equal to the quotient of two 97,389-digit integers.

Bornemann et al. wrote an appendix called "Extreme Digit-Hunting", which reported their super-accurate results in this format:

```
0.32336 74316 77778 76139 93700 <<9950 digits>> 42382 81998 70848 26513 96587 27
```

The listing goes to digit 10,002, so that if you added up the ten numbers, you'd be confident that the 10,000-th digit of the sum would be correct.

But Challenge Problem 3 proved intractable. (This was to determine what is called the "2-norm" of a matrix with infinitely many rows and columns, with entries $a_{11} = 1$, $a_{12} = 1/2$, $a_{21} = 1/3$, $a_{13} = 1/4$, $a_{22} = 1/5$, $a_{31} = 1/6, \ldots$.) With a month of computer time the authors computed 273 digits:

```
1.2742 24152 82122 81882 12340 <<220 digits>> 75880 55894 38735 33138 75269 029
```

And that's where Trefethen's constant is stuck as of today, at 273 digits.

I decided at age 20 to devote my career to number crunching, and it has given me unending satisfaction since then. My knowledge and confidence have advanced with the years, and so have the computers and software tools available. What a feeling, to be working on algorithms related to those that control spacecraft, design integrated circuits, and run satellite navigation systems — yet still be so close to elegant mathematics!

The field of numerical analysis can be defined like this:

> *Numerical analysis is the study of algorithms*
> *for the problems of continuous mathematics.*

This means problems involving real or complex variables, not just integers. "Continuous" is the opposite of "discrete", and algorithms for discrete problems have quite a different flavor and a different community of experts. Like any scientific field, numerical analysis stretches over a pure–applied range, with some people spending most of their time inventing algorithms or applying them to scientific problems, while others are more interested in rigorous analysis of their properties. In earlier centuries the leading pure mathematicians were the same people as the leading numerical ones, like Newton and Euler and Gauss, but mathematics has grown a lot since then, and now the two groups are rather separate. If you look at numbers of specialists, numerical analysis is nowadays one of the biggest branches of mathematics.

Let's finish with another ten digit problem from my file. Suppose you have three identical regular tetrahedra, each of volume 1. What's the volume of the smallest sphere you can fit them inside?

Every problem is different. This is the only one so far for which I've found myself playing with cardboard models! I made three tetrahedra, then juggled and jiggled till I thought I knew roughly what the shape of the optimal configuration must be. By numerically minimizing a function whose derivation took hours of tricky trigonometry, I got the estimate

$$22.113167462973\ldots.$$

Now by a curious coincidence, my computer tells me

$$256\pi(\sqrt{12} - \sqrt{10})^3 = 22.113167462973\ldots.$$

Have we stumbled upon the exact answer? I think so, but I'm not sure, and I certainly don't have a proof. And what in the world gave me the idea of calculating $256\pi(\sqrt{12} - \sqrt{10})^3$?[1]

6 Epilogue

Stop Press! We've had an unexpected development on Problem 1, the two cubes. I showed a draft of this article to Prof. Bengt Fornberg of the University of Colorado, one of the best numerical problem-solvers in the world. He got hooked. The problem is so simple, yet so devilishly hard!

Working with pencil and paper and the symbolic computing system Mathematica, Fornberg managed to shrink the dimensionality from six to five, then four. Then three, then two. That is, he managed to reduce Problem 1 to a two-dimensional integral to be evaluated numerically. As he peeled off one dimension after another, the formulas kept getting more complicated, and he fought hard to keep the complexity under control.

Then one morning Fornberg reported he was down to one dimension. This meant that the problem could be five-sixths solved analytically, leaving just a one-dimensional integral to be evaluated numerically. We were startled.

The next morning, Fornberg had the exact solution! It was preposterously long and complicated. He kept working, making more and more simplifications of trigonometric functions and logarithms and hyperbolic functions and their inverses, combining some terms together and also splitting some terms into two to make the result more elementary. And here is what he found:

$$F = \frac{1}{3}\left(\frac{26\pi}{3} - 14 + 2\sqrt{2} - 4\sqrt{3} + 10\sqrt{5} - 2\sqrt{6} + 26\log(2) - \log(25)\right.$$
$$+ 10\log(1 + \sqrt{2}) + 20\log(1 + \sqrt{3}) - 35\log(1 + \sqrt{5})$$
$$\left. + 6\log(1 + \sqrt{6}) - 2\log(4 + \sqrt{6}) - 22\tan^{-1}(2\sqrt{6})\right).$$

---

[1] OK, I'll tell you. My calculation led to the estimate $R \approx 0.85368706$ for the radius of the smallest sphere that can enclose three tetrahedra each of side length 1. I tried typing this number into the Inverse Symbolic Calculator at http://oldweb.cecm.sfu.ca/projects/ISC/ISCmain.html, and back came the suggestion that R might be $4(\sqrt{6} - \sqrt{5})$. Cubing this number and multiplying by $4\pi/3$ gives the volume of the sphere, and dividing that result by $\sqrt{2}/12$, which is the volume of a regular tetrahedron with side length 1, gives $256\pi(\sqrt{12} - \sqrt{10})^3$.

So now we can have as many digits as we want:

$$F \approx 0.925981260557291428093436687 0\ldots.$$

Most computational problems don't have exact solutions, but when I cook up challenges for the Problem Squad, the drive for elegance keeps me close to the edge of tractability. In this case we were lucky.

Further Reading

[1] Folkmar Bornemann, Dirk Laurie, Stan Wagon, and Jörg Waldvogel, *The SIAM 100-Digit Challenge: A Study in High-Accuracy Numerical Computing*. SIAM, Philadelphia (2004)
[2] Jonathan M. Borwein and David H. Bailey, *Mathematics by Experiment: Plausible Reasoning in the 21st Century*. AK Peters, Natick (2003)
[3] W. Timothy Gowers, June Barrow-Green, and Imre Leader (editors), *The Princeton Companion to Mathematics*. Princeton University Press, Princeton (2008)
[4] T. Wynn Tee and Lloyd N. Trefethen, A rational spectral collocation method with adaptively transformed Chebyshev grid points. *SIAM Journal of Scientific Computing* **28**, 1798–1811 (2006)
[5] Lloyd N. Trefethen, Ten digit algorithms. Numerical Analysis Technical Report NA-05/13, Oxford University Computing Laboratory. http://www.comlab.ox.ac.uk/oucl/publications/natr/

The Ever-Elusive Blowup in the Mathematical Description of Fluids

Robert M. Kerr and Marcel Oliver

Abstract. This contribution introduces you to the Euler equations of ideal fluids and the Navier–Stokes equations which govern fully developed turbulent flows. We describe some of the unresolved mathematical issues, including the "Navier–Stokes millennium problem", and the role numerical simulations play in developing this field.

1 Introduction

When you think of turbulence, you might recall the jostling and vibrations during a recent flight. Or maybe the irregular turning and twisting motions surrounding a hurricane or storm. But you don't need to look far to feel, if not see, a turbulent flow. The truth is that turbulence surrounds us almost all the time. It explains how heat and cold can quickly fill the room you are in, even if breezes are kept out. Turbulence around wings is essential for explaining the flight of airplanes and gliders. And this mundane turbulence is even less understood than the large vortices and waves primarily responsible for clear air turbulence during flight or the strong shears during storms.

Part of the reason is that, while the equations we use to represent and simulate fluids have been known for almost 200 years, we do not know whether they meet fundamental mathematical criteria. In particular,

---

Robert M. Kerr
Department of Mathematics, School of Engineering, and Centre for Scientific Computing, University of Warwick, Gibbet Hill Road, Coventry CV4 7AL UK.
e-mail: Robert.Kerr@warwick.ac.uk

Marcel Oliver
School of Engineering and Science, Jacobs University, 28759 Bremen, Germany.
e-mail: oliver@member.ams.org

could solutions to these equations develop discontinuities or singularities? If they do, our description of small scale flow must be missing essential physics.

This contribution aims at introducing the underlying mathematical problem in simple, yet precise terms and link it to insight that could be gained from computer simulation. We begin by introducing the incompressible Euler and Navier–Stokes equations of fluid dynamics. Section 3 explains conservation laws which provide important structural information. Section 4 introduces the open mathematical question of global regularity of solutions, while Section 5 sketches some heuristics which shape our current understanding. Sections 6 and 7 look at the interplay between theory and numerical experiment for guiding our choice of initial conditions and validating the results of a simulation. The final Section 8 looks into the future, sketching out new directions for research.

The article is supplemented by three more technical appendices. For readers who might not be comfortable with multivariable calculus, Appendix A introduces the main concepts and formulas in an intuitive, yet concise fashion. Appendix B explains so-called energy estimates, which give a caricature of what is known about the Navier–Stokes equations. Finally, Appendix C sketches some of the concepts behind spectral and pseudo-spectral numerical methods.

Giving complete and proper attribution is beyond what we can hope to achieve in the format of this contribution. Thus, we make no serious attempt to cite original research papers, but hope that the enterprising reader will look at the excellent recent review articles [2, 4, 5, 6, 9, 10, 12, 15, 17] and, from there, venture further into the vast body of specialized literature.

2 The Equations of Fluid Mechanics

Equations governing fluid motion may, at first glance, look intimidating. The underlying principles, however, are surprisingly simple: we apply Newton's second law of mechanics in a continuum setting and make assumptions on the mechanical forces that characterize a fluid.

Newton's equation $\boldsymbol{F} = m\boldsymbol{a}$, which says that the mechanical force on a point-particle equals its mass times its acceleration, will look familiar to anyone with some background in high school physics. Using calculus, we introduce the instantaneous velocity $\boldsymbol{v}(t)$ as the time derivative of position $\boldsymbol{x}(t)$ and the instantaneous acceleration $\boldsymbol{a}(t)$ as the time derivative of velocity $\boldsymbol{v}(t)$. When the force is a known function of position, Newton's law leads to the differential equation $\boldsymbol{F}(\boldsymbol{x}(t)) = m\,\mathrm{d}^2\boldsymbol{x}(t)/\mathrm{d}t^2$, where the particle trajectory $\boldsymbol{x}(t)$ is an unknown function to be solved for.

The Ever-Elusive Blowup in the Mathematical Description of Fluids 139

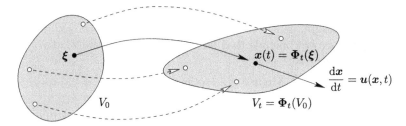

Fig. 1. The flow map $\boldsymbol{\Phi}_t$ maps the initial configuration of the fluid to its configuration at a later time $t > 0$, thereby deforming an original sub-volume V_0 into V_t. A distinguished "fluid particle" at initial location $\boldsymbol{\xi}$ is transported to location $\boldsymbol{x}(t) = \boldsymbol{\Phi}_t(\boldsymbol{\xi})$ where it moves with velocity $\boldsymbol{u}(\boldsymbol{x}(t), t) = \mathrm{d}\boldsymbol{x}/\mathrm{d}t$.

A fluid can be seen as a continuum of particles: consider some container (or *domain*) $\Omega \subset \mathbb{R}^d$ in $d = 2$ or $d = 3$ space dimensions entirely filled with point-particles. Pick a particle at location $\boldsymbol{\xi} \in \Omega$ at time $t = 0$, and denote its trajectory by $\boldsymbol{x}(t)$. There will be exactly one such trajectory emanating from every point in Ω. Therefore, the collection of trajectories defines, for each fixed time t, a mapping $\boldsymbol{\Phi}_t$ from the domain Ω into itself; see Figure 1. This mapping is referred to as the *flow map*.

Newton's law now applies to each fluid particle; more precisely, it applies to the fluid contained in each sub-volume V_t in the limit that the size of the sub-volume goes to zero. In this limit, Newton's law equates forces per volume. In particular, mass m is replaced by *mass density* $\rho(\boldsymbol{x}, t)$, the mass per volume,[1] so that Newton's law states that $\rho \boldsymbol{a}$ equals force per volume.

The setting so far applies to any type of mechanical continuum. A fluid, in particular, is characterized by the assumption that each particle pushes its neighbors equally in every direction. Then, a single scalar quantity $p(\boldsymbol{x}, t)$, the *pressure*, describes the force per area that a particle at location $\boldsymbol{x} \in \Omega$ exerts onto all its neighbors at time t. A particle is not accelerated if its neighbors push back with equal force — it is pressure *differences* that result in acceleration. This suggests that force per volume at a point is the limit of a difference quotient,[2] the negative gradient $-\boldsymbol{\nabla} p$ of the pressure. (The gradient operator $\boldsymbol{\nabla}$ is introduced in equation (15) of Appendix A.1. Here, it is understood to

---

[1] The mass contained in each finite sub-volume V is then given by the integral of ρ over the sub-volume.

[2] Consider a small box-shaped sub-volume of fluid, say of lengths (a, b, c) in the three coordinate directions. Denote the three components of force acting on the box by F_1, F_2, and F_3. Then F_1, the x_1-component of force acting on the entire box, equals the difference of pressures at the left and right ends of the box, multiplied by bc, the area of the right and left faces. To first order, this pressure difference is $-a\, \partial p / \partial x_1$, so $F_1 \approx -abc\, \partial p / \partial x_1$, and similarly for the other coordinate directions, so that $\boldsymbol{F} \approx -abc\, \boldsymbol{\nabla} p$. In the limit $a, b, c \to 0$, the higher order corrections tend to zero at a faster rate and $-\boldsymbol{\nabla} p$ remains as an exact expression for the force per volume.

act only on the space variables $\boldsymbol{x}$.) We note that the force is directed toward areas of low pressure, hence the minus sign. Equating our two expressions for force per volume, we conclude that Newton's law for a fluid reads

$$-\boldsymbol{\nabla} p(\boldsymbol{x}(t),t) = \rho(\boldsymbol{x}(t),t)\,\frac{\mathrm{d}^2\boldsymbol{x}(t)}{\mathrm{d}t^2}\,. \tag{1}$$

In principle, this is the equation we want. However, it is not quite useful yet because it mixes so-called *Eulerian quantities* and *Lagrangian quantities*. Eulerian quantities are properties of the fluid which are functions of the current position $\boldsymbol{x}$, while Lagrangian quantities are functions of the initial location, or "particle label" $\boldsymbol{\xi}$. So pressure and density are Eulerian[3] while the particle position $\boldsymbol{x}$ itself, the velocity $\mathrm{d}\boldsymbol{x}/\mathrm{d}t$, and the acceleration $\mathrm{d}^2\boldsymbol{x}/\mathrm{d}t^2$ are Lagrangian. For most purposes it is much more convenient to re-express (1) in terms of all-Eulerian quantities: writing $\boldsymbol{u}(\boldsymbol{x},t)$ to denote the velocity felt by a *stationary* observer at location $\boldsymbol{x}$ and time t, we observe that

$$\frac{\mathrm{d}\boldsymbol{x}(t)}{\mathrm{d}t} = \boldsymbol{u}(\boldsymbol{x}(t),t)\,. \tag{2}$$

Differentiating in time and using the chain rule of multivariable calculus (17) explained in Appendix A.1, we obtain a fully Eulerian expression for the acceleration of the particle,

$$\frac{\mathrm{d}^2\boldsymbol{x}(t)}{\mathrm{d}t^2} = \frac{\partial\boldsymbol{u}}{\partial t}(\boldsymbol{x}(t),t) + \boldsymbol{u}(\boldsymbol{x}(t),t)\cdot\boldsymbol{\nabla}\boldsymbol{u}(\boldsymbol{x}(t),t)\,. \tag{3}$$

Inserting (3) into (1) and dropping all arguments for simplicity, we arrive at a form of Newton's law which was first introduced by Euler in 1757:

$$\rho\left(\frac{\partial\boldsymbol{u}}{\partial t} + \boldsymbol{u}\cdot\boldsymbol{\nabla}\boldsymbol{u}\right) + \boldsymbol{\nabla}p = 0\,. \tag{4}$$

A fluid governed by (4) is called *ideal*: the model neglects possible frictional forces which can turn kinetic energy into heat and other effects caused by the molecular structure of a real fluid.

At this point, we have more unknown functions than we have equations (the d components of $\boldsymbol{u}$, pressure p, and density ρ are unknowns, but (4) provides only d equations). So we need more information coming from physics. The question is essentially this: what happens to the density when the pressure changes? There is no general answer as a gas will behave differently than, e.g., water. In this contribution, we focus on the case where the fluid is incompressible: the volume of arbitrary parcels pushed around by the flow

---

[3] In (1), the Eulerian quantities $\boldsymbol{\nabla}p$ and ρ are evaluated at the current position $\boldsymbol{x}(t)$ of a Lagrangian particle, so they are read as Lagrangian quantities. The gradient, however, must be computed with respect to Eulerian position coordinates $\boldsymbol{x}$ rather than Lagrangian labels $\boldsymbol{\xi}$.

(such as V_t in Figure 1) does not change over time.[4] The relative rate of change of a parcel volume, in the limit of vanishing size, is measured by a quantity called the divergence of the velocity field $\boldsymbol{u}$, see Appendix A.2. In particular, the flow of $\boldsymbol{u}$ is incompressible if and only if div $\boldsymbol{u} = 0$.

For simplicity, we also assume that the fluid is *homogeneous*: fluid parcels not only maintain their volume as they are pushed around by the flow, but have constant density throughout the fluid domain.[5] Then, with appropriate normalization, we can take $\rho \equiv 1$. What results are the *Euler equations* for a homogeneous incompressible ideal fluid,

$$\frac{\partial \boldsymbol{u}}{\partial t} + \boldsymbol{u} \cdot \boldsymbol{\nabla} \boldsymbol{u} + \boldsymbol{\nabla} p = 0 \,, \tag{5a}$$

$$\operatorname{div} \boldsymbol{u} = 0 \,. \tag{5b}$$

The pressure in incompressible flow is determined solely by the condition that each sub-volume must move consistently with the motion of all of its neighbors. The resulting pressure force generates the necessary instantaneous adjustment across the entire fluid domain.

So far, we have neglected friction. Due to its fundamental theoretical and practical implications, we shall look at friction in more detail. Frictional forces enhance the local coherence of the flow, i.e. they counteract, at each point, the deviation of the velocity field from its local average: if a particle moves faster than the average of its neighbors, then friction slows it down. The deviation of a function at a point from its average value on small surrounding spheres is measured by the negative of the *Laplacian* Δ, a differential operator explained in Appendix A.3, so that frictional forces should be proportional to $\Delta \boldsymbol{u}$. Adding such a term to the Euler equations (5), we obtain the *Navier–Stokes equations* for a homogeneous incompressible fluid,

$$\frac{\partial \boldsymbol{u}}{\partial t} + \boldsymbol{u} \cdot \boldsymbol{\nabla} \boldsymbol{u} + \boldsymbol{\nabla} p = \nu \, \Delta \boldsymbol{u} \,, \tag{6a}$$

$$\operatorname{div} \boldsymbol{u} = 0 \,. \tag{6b}$$

---

[4] There are certain physical effects that can only be described by a compressible model, such as acoustic waves, shock formation, and supersonic flows. If these are important, the model must be augmented with the appropriate laws of thermodynamics; in this more general case, the pressure forces are due to local imbalances in the internal energy or temperature. However, the usual state of the flow of water and the macroscopic motion of air are well described as being incompressible. There is also a computational motivation for considering incompressible flows: the effects of compressibility usually take place on much smaller scales than the motion of the bulk. These are expensive to resolve properly. Moreover, in situations where compressible effects are physically negligible, simulations of the compressible model which do not properly resolve such small scales may be very "unstable" so that it is better to start with an incompressible model right away.

[5] This is a rather mild restriction as the case of non-constant density incompressible flow is mathematically very similar to the special case considered here.

The constant $\nu > 0$ is called *coefficient of viscosity* and describes the strength of the viscous forces; it is much larger for honey, for example, than it is for water. Note that the same right hand term, $\nu \Delta \boldsymbol{u}$, appears for the same reason when modeling the flow of heat or the diffusion of a chemical; one such example is described by L.N. Trefethen in this volume [18].

The partial differential equations (5) and (6) are examples of *initial-boundary value problems*: this means that, in order to determine the flow completely, we need to specify both *initial* and *boundary values* to compute the velocity field for later times $t > 0$. For the purpose of this discussion, we assume *periodic boundary conditions* on a box-shaped domain Ω: we can think of tiling the whole of $\mathbb{R}^d$ with an array of exact copies of our domain Ω, thereby matching corresponding points at opposite faces of the box. Thus, what flows out one face of Ω appears to come back in at the opposite face.

For more physically realistic boundaries, one might alternatively specify the fluxes across the boundary — the amount of fluid moving in or out of the domain. For a Navier–Stokes flow, momentum fluxes due to the frictional forces on the boundaries are also required; this holds true even in situations where the energy dissipation due to friction is negligible — a crucial difference which is necessary for explaining lift on an aircraft wing. Here, however, we shall not consider boundary issues further; we take the point of view that fluid dynamics with periodic boundary condition is prototypical for fluid flow far away from real boundaries.

3 Conservation Laws

In classical mechanics, there are three fundamental conserved quantities: momentum, energy, and angular momentum. Each has a fluid analogue which provides important structural information. Conservation of momentum is the essence of Newton's second law and therefore already part of the picture. The kinetic energy of a point-particle is given by the expression $E = \frac{1}{2} m |\boldsymbol{v}|^2$. In a continuum, as we replace mass m by the mass density ρ, we correspondingly replace E by the kinetic energy density $\frac{1}{2} \rho |\boldsymbol{u}|^2$. For incompressible flow, all energy is kinetic, so that the total energy is obtained by integrating the kinetic energy density over the fluid domain. Here, with $\rho = 1$, the total energy reads

$$E = \frac{1}{2} \int_\Omega |\boldsymbol{u}|^2 \, \mathrm{d}\boldsymbol{x} \,. \tag{7}$$

A simple computation, detailed in Appendix B.1, shows that E remains constant under the flow of the Euler equations and is decreasing for Navier–Stokes flow where friction turns kinetic energy into heat.

What remains are conservation laws related to rotation. For Newton's equations of particle mechanics, the conserved quantity is called *angular momentum*, always defined relative to a reference point. In fluid mechanics, matters are more complicated. Instead of a reference point we must consider closed curves C_t which are transported by the flow, i.e. $C_t = \Phi_t(C_0)$. Relative to any such curve, the *circulation* is defined by the line integral

$$\Gamma_t = \oint_{C_t} \boldsymbol{u} \cdot d\boldsymbol{s} \tag{8}$$

which is computed by summing up the components of the velocity field which are tangent to C_t, see Appendix A.4. Each such Γ_t is a constant of motion for the Euler equations. Moreover, due to Stokes' theorem as explained in Appendix A.4, the circulation equals the surface integral

$$\Gamma_t = \int_{S_t} \boldsymbol{\omega} \cdot d\boldsymbol{A}, \tag{9}$$

where S_t is any oriented surface, again moving with the flow, whose boundary is the curve C_t, and where $\boldsymbol{\omega}$ denotes the *vorticity*

$$\boldsymbol{\omega} = \operatorname{curl} \boldsymbol{u} = \left(\frac{\partial u_3}{\partial x_2} - \frac{\partial u_2}{\partial x_3}, \frac{\partial u_1}{\partial x_3} - \frac{\partial u_3}{\partial x_1}, \frac{\partial u_2}{\partial x_1} - \frac{\partial u_1}{\partial x_2} \right). \tag{10}$$

The i-th component of the vorticity vector can be seen as the limit circulation per unit area in the plane perpendicular to the x_i-direction. Intuitively, it measures how much a little leaf carried by the flow would spin about the i-th coordinate vector. In two space dimensions, only the third component of (10) is nonzero and vorticity can be identified with the scalar $\partial u_2/\partial x_1 - \partial u_1/\partial x_2$.

Circulation highlights a crucial difference between flows in two and in three dimensions: since "volume" in two dimensions coincides with the notion of area, incompressibility implies that the area of S_t is a constant of the motion. In the limit of arbitrarily small area of S_0, equation (9) shows that conservation of circulation implies conservation of vorticity along flow lines. In three dimensions, there is no constraint on the area of S_t under volume-preserving transformations. Hence, conservation of circulation cannot control the magnitude of the vorticity vector. This is the reason for the qualitative differences between flow in two and three dimensions, and why the two-dimensional equations cannot have singularities, but the three-dimensional equations might.

The importance of vorticity and circulation for the question of singularity formation can be understood by a simple thought experiment. Start with a balloon filled with an incompressible non-viscous fluid at rest and tie a lasso around its waist which is contracted to a point in a finite time, forcing the fluid into two lobes above and below its waist as in Figure 2. Clearly, we have created a topological singularity as we close off the two lobes from each other at a point. Velocity and vorticity of the fluid, however, remain bounded.

Now repeat the thought experiment with a fluid inside the balloon rotating about its axis. In this example, the velocity is always tangent to the lasso so that (per the definition of the line integral) the circulation is obtained by

Fig. 2. Illustration of the rotating balloon thought experiment. As the lasso contracts, conservation of circulation implies that the velocity along the waist increases (indicated by red colors) while the velocity on the outer lobes decreases (indicated by blue colors).

integrating the magnitude of the velocity along a circle on the surface of the balloon, which is given by the product of fluid speed and the circumference of the opening. Conservation of circulation demands that both fluid speed and vorticity become infinite near the lasso as it closes, see Figure 2. This is much like an ice-skater who turns faster and faster as she moves all her mass towards the axis of her pirouette, propelled just by conservation of angular momentum. This scenario creates a much stronger singularity than the first.

The blowup question for fluids asks if any of these (or possibly other) scenarios of singularity formation may be the result of the action of a flow onto itself. For the Euler equations, it is known that any singularity is necessarily a singularity in the vorticity; this is discussed with more detail in Section 7. So the squeezing off of the rotating balloon could possibly be the movement of a sub-volume in a singular flow, while the squeezing off of the nonrotating balloon can never occur as the most singular event in the interior of a flow. For the Navier–Stokes equations, friction would prevent the velocity from becoming singular in these simple thought experiments, but in general, the question remains open, as we explain next.

4 The Clay Millennium Problem

For a mathematician, the first question when studying partial differential equations like (5) or (6) is their *well-posedness*: (i) existence of solutions — the physical system must have a way to evolve into the future, (ii) uniqueness — there must not be arbitrary choices for the evolution, and (iii) continuous dependence on the initial state — any future state is determined, to arbitrary finite precision, by the initial conditions to a sufficient finite precision.

For the incompressible Euler and Navier–Stokes equations, a complete answer to these questions is open. What is known is that both are *locally well-posed*: solutions starting out from smooth (infinitely differentiable) initial data are unique, depend continuously on the initial data, and remain smooth for at least a finite, possibly short, interval of time. Proofs of local well-posedness can be achieved by formulating the equations as a fixed point problem in a suitable space X of functions. Doing so discards much of the problem-specific structure, but we can hope to continue the solution for all times by noting that this local existence argument permits only one of the following alternatives: either the solution exists on the entire interval $[0,\infty)$ of times, or the solution exists only on some finite interval $[0,T^*)$ of times and the size of the solution, called the "norm in X", diverges as $t \to T^*$. In the latter instance, we say colloquially that the solution is "blowing up at time T^*". Hence, proving global well-posedness reduces to finding a bound on the norm in X for every $t > 0$.

The question of whether arbitrary smooth solutions of the incompressible Navier–Stokes equations in three space dimensions can be continued globally in time in this manner is now one of the seven Millennium Prize Problems posed by the Clay Mathematics Institute. It is stated as follows [8]. Either *prove that initially smooth solutions with periodic boundary conditions (or in $\mathbb{R}^3$ with strong decay conditions toward infinity) remain smooth for all times*, or *find at least one solution which blows up in finite time*. Global well-posedness for the three-dimensional incompressible Euler equations remains equally open, but is not covered by the Clay prize question.

Let us sketch some partial results. For the Euler equations in two space dimensions, we have already argued that vorticity is conserved as a scalar along flow lines. This is sufficient to prevent blowup of finite energy solutions. For Navier–Stokes in two dimensions, the dissipation of energy due to friction is sufficiently strong so that the same conclusion can also be derived via "energy estimates" which are described in more detail in Appendix B.

Even in three dimensions, solutions to the Navier–Stokes equations with general initial data can be continued past the time of their first possible singularities as "weak" or "generalized" solutions.[6] Weak solutions exist globally in time; however, it is physically troublesome that they are not known to be unique. For the three-dimensional Euler equations, only special classes of weak solutions are known; there are also examples of non-uniqueness.

It is further known that "small" solutions of the Navier–Stokes equations do not blow up. Much effort has been spent on characterizing smallness, e.g. in terms of the smallness of the initial data, of the viscosity being large, or of the solution being in some sense close to some known special solution or symmetry. Physically, all such cases can be characterized as being non-turbulent: diffusion $\nu \Delta \boldsymbol{u}$ is, in some specific sense, so strong that any perturbation coming from the $\boldsymbol{u} \cdot \boldsymbol{\nabla} \boldsymbol{u}$ term is damped away before it could lead to singularities. Intuitively, if a fluid equation for water is in danger of developing singularities, we replace the water by honey, and if the honey is sufficiently viscous, no singularities can develop. Obviously, such results are not avail-

---

[6] Although such a solution may be discontinuous or even singular, averages of the solution over small finite sub-volumes can still depend continuously on the initial state. Interpreted this way, condition (iii) in the notion of well-posedness may still be satisfied.

able for the Euler equations where $\nu = 0$. Another class of results known to hold for Navier–Stokes, but not for Euler, are so-called partial regularity results which, based on more subtle measure-theoretic arguments, state that the space-time set of singular points of weak solutions is in a certain sense small.

As a further surprise, there are known blowup solutions to both Euler and Navier–Stokes equations on unbounded domains in three as well as two dimensions. However, even before blowup, their kinetic energy per volume is also unbounded. Unbounded local energy implies, in particular, that arbitrarily large velocities arise, which cannot happen in a real physical system. Such solutions are also not covered by the Clay prize question.

5 To Blow Up or Not To Blow Up?

Let us consider a few inconclusive arguments which shape our beliefs in whether or not solutions to the Euler or Navier–Stokes equations blow up. An often cited but potentially misleading analogy are Burgers' equations, taken either with or without viscosity. These equations are obtained, respectively, from the Navier–Stokes and Euler equations by setting $p = 0$ and dropping the incompressibility constraint. These equations are not a physical model, but are of theoretical interest as a clear black-and-white picture emerges: without viscosity, each particle keeps its initial velocity and blowup occurs in the form of particle collisions. With viscosity, the global maximum of velocity remains nonincreasing and friction is strong enough to prevent blowup [12].

There is no reason to expect that the same pattern, namely, that blowup occurs if and only if viscosity is absent, pertains to real fluids. The two systems are different in very fundamental ways: for real fluid flow, there is no mechanism which can give us direct control on the magnitude of the velocity field; the difficulty arises from the coupling between transport, an inherently local phenomenon, with pressure forces which are due to the global interaction of all fluid particles. So one might think that Euler and Navier–Stokes behave worse than Burgers' equations. However, there is clear evidence that pressure stabilizes incompressible flow to some extent. In two dimensions, as we recall, we can now control the magnitude of vorticity which implies the existence of global regular solutions with or without viscosity. In three dimensions, as this control is lost, all bets are off.

Despite this, many believe that either, or both, Euler or Navier–Stokes flows develop singularities. Why? One reason is related to the *cumulative energy dissipation*

$$\nu \int_0^t \int_\Omega |\boldsymbol{\nabla} \boldsymbol{u}|^2 \, d\boldsymbol{x} \, dt \, . \tag{11}$$

This quantity measures how much energy has been lost up to time t, as is explicitly shown in Appendix B.1. We would expect that, over a fixed interval of time $[0,t]$, this quantity is continuous as a function of ν. However, there is strong numerical and experimental evidence that the cumulative energy dissipation does not converge to zero as $\nu \to 0$. This phenomenon is referred to as *anomalous dissipation* and is assumed in most attempts to model turbulence [7]. Mathematically, anomalous dissipation implies loss of smoothness at least for the Euler equations. Its implications for the Navier–Stokes equations are less clear.

Suppose, for the sake of argument, that Euler solutions do blow up, but Navier–Stokes solutions don't. Then regularity for the Navier–Stokes equations should come from the viscous term. However, the known mechanisms in which viscosity acts do not suffice to prove that viscosity could always control the nonlinearity. (For example, in L.N. Trefethen's model blowup problem [18], the diffusive term is insufficiently strong to prevent blowup.) So unless there is a yet unknown magical mechanism, at least some Navier–Stokes solutions might blow up, too.

So why do others believe that the Euler and Navier–Stokes equations do not have singularities? There are two reasons. First, because the numerical evidence remains, despite much effort, inconclusive. There are simulations on both sides of the argument [3, 11, 13, 15] none of which, however, establishes a "road to blowup" as is known for other models. Second, because once the continuum description of fluids is accepted, there is nothing obviously missing or incomplete. This is bolstered by the success of the Navier–Stokes equations as a deterministic theory when compared with almost every piece of experimental evidence.

6 Collapse of Vortex Tubes

Let us now look more closely at the role that simulation might play in solving the blowup problem. As with experiments in a real-world laboratory, before embarking on any sort of computational experimentation, one must first identify computational scenarios which would provide the most insight within the hardware and algorithmic constraints under which we must operate.

Our objective is to find configurations which quickly develop localized intense dynamics. Such scenarios can then be probed for signs of singularity formation or for signs of depletion of the nonlinear self-amplification. Initialization with random data was used first and indicated that intense events tend to occur in tube-like structure that rotate about their axis of symmetry. They are referred to as *vortex tubes* and occur in many natural flows, including tornadoes descending from strongly rotating storms or around the rising parcels of heated air in afternoon thunder clouds. Vortex tubes are often not easily visible, but can be visualized through condensation due to the low pressure in the vortex cores, air bubbles sucked into the cores, or injection of dye.

Fig. 3. Self-propagating steam rings ejected from an eruption at the south east crater of the volcano Etna in 2000. Notice how the shadow of the ring travels across the slope. Photos by Juerg Alean from http://www.swisseduc.ch/stromboli/etna/etna00/etna0002photovideo-en.html?id=4.

Simulations and data from experiments indicate two further trends. Vortices in turbulent flows amplify by stretching, much as in the rotating balloon thought experiment in Section 3. And the most intense events, if not the most frequent, tend to involve pairs of parallel counter-rotating ("antiparallel") vortex tubes which initially self-propagate. This is why many studies of singularity formation focus on such pairs [15]. There are also high resolution studies with smooth, highly symmetric initial conditions which might be showing signs of similar singular behavior [11]. Simulations with random data continue to play a crucial role in the study of turbulence, but are now considered too "noisy" to reveal the local structure of possible singularity formation.

In a flow without viscosity, two exactly linear, antiparallel vortex tubes will simply propagate at constant velocity. Similarly, vortex rings, which can appear as smoke rings, are a beautiful example of self-propagation: diagonally across the core, the direction of vorticity is anti-aligned, pushing fluid through the center and dragging the vortex ring with it; see Figure 3. Another good example of self-propagation and subsequent break-down are the vortices shed by aircraft wings. Sometimes they are visible as condensation trails, where water vapor condenses due to pressure and temperature drops in the vortex cores; they can also be made visible by smoke generators, as in the NASA study shown in Figure 4. (Typical high-altitude aircraft condensation trails come from the outflow of jet engines, but eventually they are engulfed by the wing-tip vortices.) Further downstream, these tube-like structures are twisting and starting to attract one another. This is known as the Crow in-

Fig. 4. This picture shows the wing-tip vortices which are associated with the circulation generating lift coming off the wings of a Boeing 727 aircraft. They were made visible by smoke generators installed on the tips of the aircraft wings. (NASA photograph number ECN-3831.)

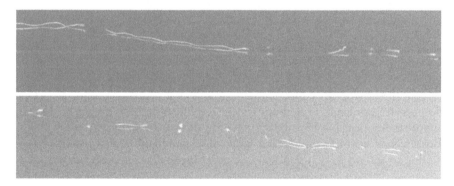

Fig. 5. Breakdown via the so-called Crow instability of a pair of vortex tubes trailing an aircraft. What is seen is the interaction of the vortices with the condensation trail from the jet engines; the lower half of the picture continues the upper half on the right. From http://commons.wikimedia.org/wiki/File:The_Crow_Instability.jpg.

stability; see Figure 5. Eventually, the tubes will touch and reconnect, before becoming turbulent and disappearing.

Why are the two vortex tubes attracted to one another? When an exactly linear self-propagating pair of vortex tubes is perturbed, the tubes will be stretched somewhere along their axis. Under incompressibility, this must be compensated for by compression in the perpendicular directions. The tubes become longer and thinner, much as when chewing gum is pulled, and move closer together; see Figure 6. This mechanism is self-amplifying and leads to a rapid generation of small scale structures, as in Figure 7. Many measures of the complexity of the flow, especially vorticity and pressure gradients, grow

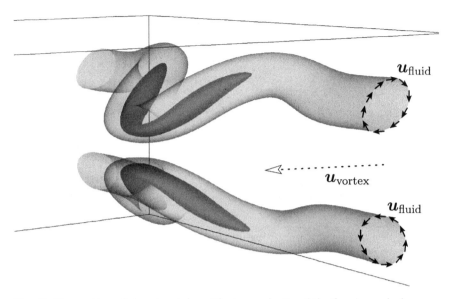

Fig. 6. Two antiparallel vortex tubes. The true velocity of the flow is marked u_{fluid}; the apparent direction of propagation of the vortex structure is indicated as u_{vortex}. The colored surfaces are surfaces of constant vorticity modulus at 60% and 90% of peak vorticity. From [3].

around these structures. One of the questions that remain unanswered is this: do pressure gradients contribute to the amplification of vorticity, or do they suppress any possible blowup by repelling the vortices and flattening them? Both trends have been observed, depending upon the initial conditions or, adding to the mystery, upon what stage of the evolution is considered.

7 Numerical Error

In areas such as number theory or discrete mathematics, computers can find examples or counterexamples; sometimes it is even possible to achieve computer-assisted proofs. The space-time continuum of fluid motion, however, can only be approximated on a discrete computational mesh by finitely many floating point values. Consequently, computers are fundamentally incapable of *proving* that solutions to the Euler or Navier–Stokes problems are well-behaved. On the other hand, *knowing* that the solutions are well-behaved, we can prove that, given sufficient computational resources, we could numerically solve the equation to any required accuracy. Yet, there are many ways how computations interact with mathematical analysis to provide us with a better understanding of the blowup question. Given the

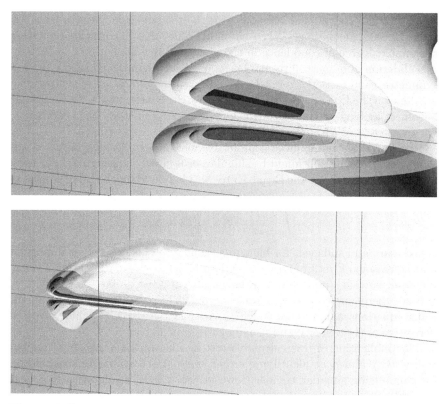

Fig. 7. Collapse of two antiparallel vortex tubes. Snapshots at $t = 5.6$ and $t = 8.1$ with a conjectured time of singularity $T^* \approx 11$. From [3].

proper configurations, simulations can inspire conjectures, validate assumptions, and probe the properties of inequalities.

A large number of numerical approximation methods ("schemes") for fluid equations have been devised, each with distinct advantages and disadvantages. For example, many engineering problems use adaptive schemes that refine the computational mesh locally in regions of interest. Sometimes it is also possible to choose an initial condition which makes optimal use of a fixed computational mesh, and then use simple and fast numerics, such as spectral methods as detailed in Appendix C.

No matter which approach is used, a near-singular flow will develop small scale features which cannot be represented well with a fixed and finite number of degrees of freedom. As the flow develops, errors will grow, so that we need some means of validating the accuracy of the computation. The most complete validation possible would be to refine on progressively finer meshes, requiring progressively larger computers. Since this is impossible from a practical point of view, the usual compromise is to monitor the preservation of invariants and check for anomalous growth of small scale features for a few refined mesh calculations. When carefully done, this can provide us with a reasonable degree of confidence in the quality of the simulation.

Validation of numerical results becomes easier when the equations possess symmetries and conservation laws. For this reason, almost all blowup studies are done on the Euler rather than the Navier–Stokes equations. An additional benefit of looking at non-viscous flow is that one can study the geometry of vortex stretching without possible interference of viscous effects.

Besides validation, we need computable measures to decide whether a particular simulation result might be singular, or is clearly nonsingular. The most famous such criterion is the Beale–Kato–Majda bound for the Euler equations [1], which says that

$$\int_0^{T^*} \max_{\boldsymbol{x}\in\Omega} |\boldsymbol{\omega}(\boldsymbol{x},t)| \, \mathrm{d}t = \infty \qquad (12)$$

is necessary and sufficient for blowup at time T^*. This criterion is important for two reasons. First, it gives a bound on how fast singularities can develop: the peak vorticity must blow up essentially at least as fast as $(T^* - t)^{-1}$. Second, there can be no singularities in higher derivatives of velocity in the Euler equations unless there is a singularity in the vorticity, which is a first derivative of the velocity and usually easily calculable.

The publication of this calculable test in 1984 fueled a decade of numerical activity aimed at identifying either singular structures, or mechanisms for suppressing singular trends. However, no scientific consensus had been reached, with only one calculation [14] providing modest consistency with a power law singularity while maintaining sufficient resolution in all three directions. It also became apparent that the Beale–Kato–Majda criterion by itself was insufficient for discriminating between the competing claims. Several additional tests for singular behavior that are both independent and calculable were subsequently proposed [2, 5, 13, 16]. Still, whether the strongest claims for singular behavior are consistent with the mathematical bounds remains an open question [3, 13]. It now appears feasible that a consolidated effort involving adequate high performance computing resources, the latest in adaptive mesh methods, and the use of better initial perturbation profiles could either substantiate the proposed singular scaling regime, or would clearly show how the nonlinearities generate a negative feedback that suppresses singular trends.

8 An Invitation to Research

New blowup criteria — some that are more robust, as well as some that are more refined — could make a real difference. Numerically robust tests would involve space integrals or averages. One candidate might be the enstrophy, the integral over the square of the vorticity, which appears promising but needs theoretical support. In [3], for example, the Euler enstrophy appears to follow a power law consistent with the best known upper bound on the growth of enstrophy in the Navier–Stokes equations. Whether this is coincidental or whether there is a deeper connection remains a mystery. More refined blowup tests, on the other hand, might make explicit reference to the local geometry of vortex lines and vortex structures.

More generally, there remain open questions about the relationship between the Navier–Stokes and Euler equations. In the presence of boundaries, the limit of vanishing viscosity is still not well understood [2]. Further, it remains a mystery whether a global regularity result for Euler would imply one for Navier–Stokes, as one would naively expect because one would think viscosity can only dampen the development of singularities [5]. Nonetheless, we believe that a breakthrough on the Navier–Stokes problem will come via a breakthrough on the Euler problem. One reason is that the Navier–Stokes viscosity is mathematically well understood, yet is insufficient to control the nonlinearity. Another reason is that the Euler equations have conservation laws, while the Navier–Stokes equations do not, which can be used to monitor the reliability of numerical simulations probing for blowup.

Independent of these hard problems, it is always worthwhile building intuition with lower dimensional toy problems which share some similarity with the three-dimensional Navier–Stokes and Euler equations. In some cases, simulations have been used to predict existence and non-existence of singularities. The extreme situations created to address these issues, often found after a painful period of numerical experimentation, continue to inspire new mathematics which is then used to validate numerical predictions.

In practical applications, one can often "model" the effects of small scales which cannot be resolved computationally. In "large eddy simulations", for example, the Navier–Stokes viscous term is replaced by an eddy viscosity designed to represent the average effects of viscosity over a computational cell. In a global weather calculation, almost everything is modeled. Such approaches are often successful in preserving crucial statistical properties of the solution; their importance cannot be overestimated, yet their mathematical study remains wide open. Especially needed is the development of new mathematical concepts to address the relation of accuracy at large scales to a probabilistic notion of accuracy at small scales.

Mathematical fluid dynamics and, more generally, partial differential equations is a field where analysis, physics, and computation meet and frequently progress jointly. It is a field where deep mathematical questions and application driven problems sit side by side. And it is a field which, at this point in its long history, is as vibrant as ever.

Appendix A. A Brief Guided Tour of Vector Calculus

For functions in one variable, the derivative $f' = \mathrm{d}f/\mathrm{d}x$ denotes the local rate of change of f per unit distance, and the fundamental theorem of calculus relates these local rates to the global change of f within an interval $[a, b]$. For functions in several variables, as necessarily occur when modeling fluid flow, local rates of change are measured by directional derivatives which give rise to four important differential operators: gradient, divergence, curl, and Laplacian. These differential operators describe local properties of a function or vector field which are related to global changes via the integral theorems of Gauss and Stokes. We will explain these basic concepts without proofs and precise statements of assumptions under the premise that our readers are already familiar with single variable calculus and some analytic geometry. For further background, there are many excellent textbooks and we encourage the reader to find his or her favorite, or to search on the internet.

A.1 Directional Derivative, Gradient, and Chain Rule

Let $U \subset \mathbb{R}^n$ be an open set, $f \colon U \to \mathbb{R}$ a function, and consider a point $\boldsymbol{x} = (x_1, \ldots, x_n) \in U$. (We use boldface symbols for vector-valued variables or functions, and plain symbols for real-valued variables or functions.) We ask for the rate of change of f as we vary its argument $\boldsymbol{x}$ in some direction $\boldsymbol{v} = (v_1, \ldots, v_n) \in \mathbb{R}^n$. This question can be answered by taking the single-variable derivative of the function $t \mapsto f(\boldsymbol{x} + t\boldsymbol{v})$, which is well-defined for small values of t. Then the local rate of change,

$$\left.\frac{\mathrm{d}f(\boldsymbol{x} + t\boldsymbol{v})}{\mathrm{d}t}\right|_{t=0} = \lim_{t \to 0} \frac{f(\boldsymbol{x} + t\boldsymbol{v}) - f(\boldsymbol{x})}{t}, \tag{13}$$

if it exists, is called the *directional derivative* of f at $\boldsymbol{x}$ in the direction $\boldsymbol{v}$.[7]

When $\boldsymbol{v} = (0, \ldots, 0, 1, 0, \ldots, 0)$ is a unit vector with a single 1 in the i-th coordinate, the associated directional derivative is referred to as the i-th *partial derivative*, written $\partial f/\partial x_i$ or $\partial_i f$ for short. It is computed by taking the single-variable derivative of f with respect to x_i while holding the remaining components of $\boldsymbol{x}$ constant.

More generally, we can look at the rate of change of f as its argument changes along an arbitrary smooth curve which is parameterized by a function $\boldsymbol{\phi} \colon (a, b) \to U$. The vector components of $\boldsymbol{\phi}$ are denoted $\phi_1, \ldots, \phi_n$, so that $\boldsymbol{\phi}(t) = (\phi_1(t), \ldots, \phi_n(t))$. Then the multivariate *chain rule* asserts that

$$\frac{\mathrm{d}}{\mathrm{d}t} f(\boldsymbol{\phi}(t)) = \sum_{i=1}^{n} \frac{\mathrm{d}\phi_i(t)}{\mathrm{d}t} \left.\frac{\partial f}{\partial x_i}\right|_{\boldsymbol{x}=\boldsymbol{\phi}(t)}. \tag{14}$$

---

[7] The "vertical bar" notation used in (13) and subsequent expressions indicates that the derivative should be computed *before* the indicated argument substitution is applied.

The sum in this expression can be understood as a vector dot product between $d\boldsymbol{\phi}/dt = (d\phi_1/dt, \ldots, d\phi_n/dt)$, which can be thought of as the velocity of a point moving on the curve, and the vector of partial derivatives

$$\boldsymbol{\nabla} f \equiv (\partial f/\partial x_1, \ldots, \partial f/\partial x_n), \qquad (15)$$

which is called the *gradient* of f. Using familiar dot product notation $\boldsymbol{u} \cdot \boldsymbol{v} = u_1 v_1 + \cdots + u_n v_n$, we can then write the chain rule (14) as

$$\frac{d}{dt} f(\boldsymbol{\phi}(t)) = \frac{d\boldsymbol{\phi}(t)}{dt} \cdot \boldsymbol{\nabla} f \Big|_{\boldsymbol{x}=\boldsymbol{\phi}(t)}. \qquad (16)$$

This is like the one-dimensional chain rule $(f(\phi(t))' = \phi'(t) f'(\phi(t)))$, except we have to take the contributions in all n coordinate directions into account.

Applying the chain rule with $\boldsymbol{\phi}(t) = \boldsymbol{x} + t\boldsymbol{v}$ so that $d\boldsymbol{\phi}/dt = \boldsymbol{v}$, we find that $\boldsymbol{v} \cdot \boldsymbol{\nabla} f$ expresses the directional derivative of f in the direction $\boldsymbol{v}$. Among all vectors $\boldsymbol{v}$ of unit length, $\boldsymbol{v} \cdot \boldsymbol{\nabla} f$ is maximal when $\boldsymbol{v}$ aligns with $\boldsymbol{\nabla} f$; we conclude that the gradient $\boldsymbol{\nabla} f$ is a vector which points into the direction in which f has the greatest directional derivative at $\boldsymbol{x}$, and that the magnitude of $\boldsymbol{\nabla} f$ is the directional derivative of f in this direction. It is often convenient to think of $\boldsymbol{\nabla} = (\partial_1, \ldots, \partial_n)$ as a vector of differentiation symbols.

A *vector field* is a function $\boldsymbol{u} \colon U \to \mathbb{R}^n$ which assigns a vector to each $\boldsymbol{x} \in U$. A typical example is a fluid which has, at each point, a velocity vector $\boldsymbol{u}(\boldsymbol{x})$. When applying the directional derivative $\boldsymbol{u} \cdot \boldsymbol{\nabla}$ to a vector field $\boldsymbol{v}$, it acts on each component separately, i.e., $\boldsymbol{u} \cdot \boldsymbol{\nabla} \boldsymbol{v} = (\boldsymbol{u} \cdot \boldsymbol{\nabla} v_1, \ldots, \boldsymbol{u} \cdot \boldsymbol{\nabla} v_n)$.

In fluid dynamics, we typically encounter time dependent functions and vector fields. Hence, we must notationally distinguish the space variables $\boldsymbol{x}$ from time t. We write $\boldsymbol{x} \in \Omega \subset \mathbb{R}^d$ (usually with $d = 2$ or $d = 3$) to denote a point in our fluid domain, and (a, b) for a time interval. We use $\boldsymbol{\nabla} = (\partial_1, \ldots, \partial_d)$ to denote the gradient with respect to the space coordinates only, and ∂_t for the partial derivative with respect to time. The chain rule (14) for a function $f \colon \Omega \times (a, b) \to \mathbb{R}$ and $\boldsymbol{\psi} \colon (a, b) \to \Omega$ with $n = d + 1$ reads

$$\frac{d}{dt} f(\boldsymbol{\psi}(t), t) = \partial_t f \Big|_{\boldsymbol{x}=\boldsymbol{\psi}(t)} + \frac{d\boldsymbol{\psi}(t)}{dt} \cdot \boldsymbol{\nabla} f \Big|_{\boldsymbol{x}=\boldsymbol{\psi}(t)}. \qquad (17)$$

The first term on the right records the change of f coming directly from the time dependence of f, while the second term records the changes from moving along the curve $\boldsymbol{\psi}$. This form of the chain rule arises from (16) by setting $U = \Omega \times (a, b)$ and $\boldsymbol{\phi}(t) = (\boldsymbol{\psi}(t), t)$ so that $d\boldsymbol{\phi}(t)/dt = (d\psi_1(t)/dt, \ldots, d\psi_d(t)/dt, 1)$ and the vector of partial derivatives of f reads $(\partial_1 f, \ldots, \partial_d f, \partial_t f) = (\boldsymbol{\nabla} f, \partial_t f)$.

A.2 Source Strength and the Divergence of a Vector Field

Given a vector field $\boldsymbol{u}$, we measure its "source strength" as follows. For a given sub-volume, for instance a small box Q with boundary S, we define the *flux* of $\boldsymbol{u}$ through S as the surface integral

$$\mathrm{Fl} = \int_S \boldsymbol{u} \cdot \mathrm{d}\boldsymbol{A} \,. \tag{18}$$

This integral expresses that we are summing up the component of $\boldsymbol{u}$ which is perpendicular to the boundary S. The flux measures the net volume of fluid crossing S per unit time. Note that the component of $\boldsymbol{u}$ which is parallel to the boundary does not contribute to the flux in or out of the box.

If Q is subdivided into two subelements of volume, say Q_1 and Q_2, then the flux out of Q equals the sum of the fluxes out of Q_1 and out of Q_2, as the contributions on the common boundary cancel. By further repeated subdivision, we can localize the contributions to the flux which are generated by smaller and smaller sub-volumes. Finally, we define the *divergence* of $\boldsymbol{u}$ at $\boldsymbol{x}$, written div $\boldsymbol{u}$, as the flux out of Q divided by the volume of Q, in the limit of smaller and smaller volume elements Q. The divergence measures

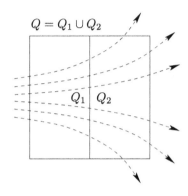

how much flux is produced per volume, and this is the "source density" or "source strength". An immediate consequence of this definition is that the flux out of Q equals the volume integral of the source strength over Q, symbolically written as

$$\int_S \boldsymbol{u} \cdot \mathrm{d}\boldsymbol{A} = \int_Q \mathrm{div}\, \boldsymbol{u}\, \mathrm{d}\boldsymbol{x}\,, \tag{19}$$

where S denotes the surface of Q. This expression is usually referred to as *Gauss' divergence theorem*. A simple calculation reveals that

$$\mathrm{div}\, \boldsymbol{u} = \partial_1 u_1 + \cdots + \partial_n u_n\,, \tag{20}$$

which can be written symbolically as $\boldsymbol{\nabla} \cdot \boldsymbol{u}$. (Usually, this equation is used as the definition of the divergence; then one needs to prove Gauss' theorem (19). Here, we have essentially taken (19) as a definition, and it is (20) that needs proof.)

The Gauss theorem implies, in particular, that when div $\boldsymbol{u}(\boldsymbol{x}) > 0$, more fluid comes out of a small box around $\boldsymbol{x}$ than flows in, and there is a "source" at $\boldsymbol{x}$ — the fluid expands. Correspondingly, if div $\boldsymbol{u}(\boldsymbol{x}) < 0$, then more fluid comes in than flows out — the fluid contracts. If div $\boldsymbol{u} = 0$ at all points, then inflow balances outflow everywhere and the flow preserves volume.

A.3 Deviations from Averages and the Laplace Operator

The *Laplacian* Δ of a real-valued function f in n variables measures how much the value of f at a point x differs from the average of f on small spheres around x: let $S_\varepsilon(x)$ denote the sphere of radius ε centered at x and let $\mathrm{Av}(f, S_\varepsilon(x))$ denote the average value of f on this sphere. We then define

$$\Delta f = 2n \lim_{\varepsilon \to 0} \frac{\mathrm{Av}(f, S_\varepsilon(x)) - f(x)}{\varepsilon^2}. \tag{21}$$

An intricate computation based on Gauss' theorem (19) shows that the Laplacian is the differential operator

$$\Delta f = \partial_1 \partial_1 f + \cdots + \partial_n \partial_n f, \tag{22}$$

which can be written symbolically as $\nabla \cdot \nabla f$. Applied to a vector field, the Laplacian acts on each component separately, i.e., $\Delta u = (\Delta u_1, \ldots, \Delta u_n)$.

To motivate the equivalence of (21) and (22), consider an affinely linear function $f(x) = ax + b$ in one variable; then $\mathrm{d}^2 f/\mathrm{d}x^2 = 0$ and $f(x) = (f(x - \varepsilon) + f(x + \varepsilon))/2$, so $f(x)$ equals the average of all values at distance ε from x; we rewrite this as $f(x + \varepsilon) + f(x - \varepsilon) - 2f(x) = 0$. This is of course not so for non-linear f, but we always have

$$\frac{\mathrm{d}^2 f}{\mathrm{d}x^2} = \lim_{\varepsilon \to 0} \frac{f(x + \varepsilon) + f(x - \varepsilon) - 2f(x)}{\varepsilon^2}, \tag{23}$$

the expression for the one-dimensional Laplacian.

A.4 Circulation and the Curl of a Vector Field

Our final differential operator, the curl, is most easily introduced for $n = 3$. There is, however, the beautiful more abstract framework of "differential forms" in which Gauss' theorem (19) and Stokes' theorem below take a simple common form.

Let $C \subset \mathbb{R}^3$ denote a smooth curve, parametrized as $s \colon [a, b] \to C$. The curve is called *closed* if $s(a) = s(b)$, and we assume that the parametrization is traversing the curve once. The *circulation* of u along C is then defined by

$$\oint_C u \cdot \mathrm{d}s = \int_a^b u(s(r)) \cdot s'(r) \, \mathrm{d}r. \tag{24}$$

(The small circle in the integral sign indicates that the curve of integration is closed.) The line integral can be thought of as summing up the components of u which are tangential to the curve; it is easy to show that it is independent of the choice of parametrization. It measures the amount of spinning of the flow along the curve C. For instance, water flow around the drain of a water basin often develops a strong circulation around the drain.

Suppose there is a piece of a surface, say S, for which C is the boundary. The circulation in (24) is found by going once around S. How does the total circulation change if we subdivide S into two subsurfaces S_1 and S_2? When computing the circulation around S_1 and S_2 separately, their common boundary curve is traversed twice, but in opposite direction. The corresponding contributions to the total circulation

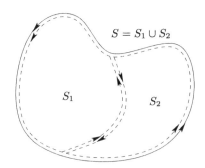

thus cancel in the sum, and only the contribution from the boundary of S remains. Again, we can ask which part of the surface is responsible for producing circulation by further subdividing S. As the area of the subdivisions tends to zero, the surfaces look more and more like planes, so that it suffices to look at the limit circulation per unit area for planar surface elements. This limit circulation per unit area is called the *curl* of $\boldsymbol{u}$ at point $\boldsymbol{x}$, and is written $\operatorname{curl}\boldsymbol{u}$. It is a vector whose component in the direction perpendicular to a plane P is the circulation of $\boldsymbol{u}$ around a small surface $S \subset P$ containing $\boldsymbol{x}$, divided by the area of S, in the limit that this area tends to zero. Consequently,

$$\int_S \operatorname{curl}\boldsymbol{u} \cdot \mathrm{d}\boldsymbol{A} = \oint_C \boldsymbol{u} \cdot \mathrm{d}\boldsymbol{s}, \qquad (25)$$

where we may interpret $\operatorname{curl}\boldsymbol{u}$ as a vector field and the left hand integral as the total flux of $\operatorname{curl}\boldsymbol{u}$ through the surface S. Equation (25) is usually referred to as *Stokes' theorem*. To determine the three vector components of the curl, it is sufficient to compute the limit circulation per unit area in the each of the three coordinate planes. A calculation which we will not reproduce here, but that we encourage our readers to find for themselves, yields

$$\operatorname{curl}\boldsymbol{u} = (\partial_2 u_3 - \partial_3 u_2, \partial_3 u_1 - \partial_1 u_3, \partial_1 u_2 - \partial_2 u_1), \qquad (26)$$

which can be written symbolically as the vector product $\boldsymbol{\nabla} \times \boldsymbol{u}$. If $\boldsymbol{u}$ is the velocity field of a fluid, we refer to $\operatorname{curl}\boldsymbol{u}$ as the *vorticity*. Its third component, for instance, measures how much the flow restricted to the (x_1, x_2)-plane rotates about the axis through $\boldsymbol{x}$ in the x_3-direction.

Appendix B. Energy Estimates

In this appendix, we derive simple estimates for smooth solutions of the Navier-Stokes equations which are remarkably close to the best known. We treat only the case of periodic boundary conditions. The main result is the "energy relation" (32), which expresses that the total kinetic energy is constant for solutions of the Euler equations and strictly decreasing for solutions of the Navier–Stokes equations where friction comes into play. We also indicate how energy estimates can lead to bounds on derivatives, and where the difficulty in establishing better bounds comes from.

B.1 Energy is Non-Increasing

We begin by taking the dot product of the Navier–Stokes momentum equation (6a) with $\boldsymbol{u}$ and integrating over the space domain Ω:

$$\int_\Omega \boldsymbol{u} \cdot \frac{\partial \boldsymbol{u}}{\partial t}\,\mathrm{d}\boldsymbol{x} + \int_\Omega \boldsymbol{u} \cdot (\boldsymbol{u} \cdot \boldsymbol{\nabla} \boldsymbol{u})\,\mathrm{d}\boldsymbol{x} + \int_\Omega \boldsymbol{u} \cdot \boldsymbol{\nabla} p\,\mathrm{d}\boldsymbol{x} = \nu \int_\Omega \boldsymbol{u} \cdot \Delta \boldsymbol{u}\,\mathrm{d}\boldsymbol{x}. \quad (27)$$

Writing

$$|\boldsymbol{u}|^2 = \sum_{i=1}^d |u_i|^2 \quad \text{and} \quad |\boldsymbol{\nabla}\boldsymbol{u}|^2 = \sum_{i,j=1}^d |\partial_i u_j|^2, \quad (28)$$

recognizing that $\partial|\boldsymbol{u}|^2/\partial t = 2\,\boldsymbol{u} \cdot \partial \boldsymbol{u}/\partial t$ and $\boldsymbol{u} \cdot \boldsymbol{\nabla}|\boldsymbol{u}|^2 = 2\,\boldsymbol{u} \cdot (\boldsymbol{u} \cdot \boldsymbol{\nabla}\boldsymbol{u})$, and moving the time derivative under the integral, we obtain

$$\frac{1}{2}\frac{\mathrm{d}}{\mathrm{d}t}\int_\Omega |\boldsymbol{u}|^2\,\mathrm{d}\boldsymbol{x} + \frac{1}{2}\int_\Omega \boldsymbol{u} \cdot \boldsymbol{\nabla}|\boldsymbol{u}|^2\,\mathrm{d}\boldsymbol{x} + \int_\Omega \boldsymbol{u} \cdot \boldsymbol{\nabla} p\,\mathrm{d}\boldsymbol{x} = \nu \int_\Omega \boldsymbol{u} \cdot \Delta \boldsymbol{u}\,\mathrm{d}\boldsymbol{x}. \quad (29)$$

Now, if f is a function and $\boldsymbol{v}$ a vector field, then $\mathrm{div}(f\boldsymbol{v}) = f\,\mathrm{div}\,\boldsymbol{v} + \boldsymbol{v} \cdot \boldsymbol{\nabla} f$. Thus, applying Gauss' theorem (19) to $f\boldsymbol{v}$ and noting that the boundary integral on the left of (19) vanishes (because contributions from opposite boundary faces of our periodic domain cancel), we obtain the multi-dimensional "integration by parts" formula

$$\int_\Omega \boldsymbol{v} \cdot \boldsymbol{\nabla} f\,\mathrm{d}\boldsymbol{x} = -\int_\Omega f\,\mathrm{div}\,\boldsymbol{v}\,\mathrm{d}\boldsymbol{x}. \quad (30)$$

Hence, the second and third terms in (29) vanish altogether: after integrating by parts, the integrands each contain the factor $\mathrm{div}\,\boldsymbol{u}$, which is zero. Writing $\Delta \boldsymbol{u} = \boldsymbol{\nabla} \cdot \boldsymbol{\nabla}\boldsymbol{u}$ and integrating by parts in the last term of (29), we find

$$\frac{1}{2}\frac{\mathrm{d}}{\mathrm{d}t}\int_\Omega |\boldsymbol{u}|^2\,\mathrm{d}\boldsymbol{x} = -\nu \int_\Omega |\boldsymbol{\nabla}\boldsymbol{u}|^2\,\mathrm{d}\boldsymbol{x}. \quad (31)$$

Integration with respect to time finally yields the "energy relation"

$$E(t) \equiv \frac{1}{2}\int_\Omega |\boldsymbol{u}(t)|^2\,\mathrm{d}\boldsymbol{x} = E(0) - \nu \int_0^t \int_\Omega |\boldsymbol{\nabla}\boldsymbol{u}(s)|^2\,\mathrm{d}\boldsymbol{x}\,\mathrm{d}s. \quad (32)$$

Since the integrand in the last term is non-negative, the energy E is non-increasing and the cumulative energy dissipation (11) is bounded by the initial energy $E(0)$. For the Euler equations where $\nu = 0$, energy is a constant of the motion, but there is no implied bound on the space-time integral of $|\nabla u|^2$.

B.2 Bounds on Derivatives

Let us now consider how the energy relation may control derivatives of u. The essential difficulty lies in getting control of, in particular, the integral of $|\nabla u|^2$ over Ω pointwise in time; if this were done, bounds on derivatives of any order would follow by standard arguments. In an attempt to prove the required bound, we take the dot product of (6a) with Δu and integrate over the spatial domain as before,

$$\int_\Omega \Delta u \cdot \partial_t u \, dx + \int_\Omega \Delta u \cdot (u \cdot \nabla u) \, dx + \int_\Omega \Delta u \cdot \nabla p \, dx = \nu \int_\Omega |\Delta u|^2 \, dx. \quad (33)$$

Integration by parts readily identifies the first term as the time derivative of $|\nabla u|^2$ and lets the pressure contribution vanish as before. However, the second term — containing the contribution from the Navier–Stokes nonlinearity — does not vanish. After a short computation, we obtain

$$\frac{1}{2} \frac{d}{dt} \int_\Omega |\nabla u|^2 \, dx + \sum_{i,j,k=1}^d \int_\Omega \partial_i u_j \, \partial_i u_k \, \partial_k u_j \, dx = -\nu \int_\Omega |\Delta u|^2 \, dx. \quad (34)$$

The second term looks complicated and does not have a definite sign. Simple-mindedly, we bound each gradient by its Euclidean length: when $\nu > 0$,

$$\left| \sum_{i,j,k=1}^d \int_\Omega \partial_i u_j \, \partial_i u_k \, \partial_k u_j \, dx \right| \leq \int_\Omega |\nabla u|^3 \, dx$$

$$\leq c_1 \left(\int_\Omega |\Delta u|^2 \, dx \right)^{\frac{d}{4}} \left(\int_\Omega |\nabla u|^2 \, dx \right)^{\frac{6-d}{4}}$$

$$\leq \nu \int_\Omega |\Delta u|^2 \, dx + c_2 \left(\int_\Omega |\nabla u|^2 \, dx \right)^d, \quad (35)$$

where c_1 and $c_2 = c_2(\nu)$ are known positive constants and $d = 2$ or $d = 3$. The proof of the second inequality requires some technical tricks in multivariate integration; the values of the right hand exponents, however, follow from a simple scaling argument which expresses that the inequality preserves physical units. The third inequality is simply a variant of the arithmetic-geometric mean inequality. Altogether,

$$\frac{1}{2} \frac{d}{dt} \int_\Omega |\nabla u|^2 \, dx \leq c_2 \left(\int_\Omega |\nabla u|^2 \, dx \right)^d. \quad (36)$$

When $d = 2$ and $\nu > 0$, equation (36) can be interpreted as a linear nonautonomous differential inequality: due to the boundedness of the cumulative energy dissipation (11), a standard "integrating factor" argument yields a global bound on the integral of $|\nabla u|^2$. When $d = 3$, this differential inequality is truly nonlinear so that the implied bound blows up in finite time. In in other words, energy and cumulative energy dissipation are too weak to control the development of fine scales.

The argument can be tweaked to show that when $d = 3$, global solutions exist provided the initial data is sufficiently small, the viscosity ν is sufficiently large, or the initial data is in various ways close to some global regular special situation (as already mentioned in Section 4). Tweaking at this level or looking for more clever choices of function spaces, however, can neither alter the dimensional scaling of the terms in the equation nor the fact that in three dimensions the energy relation provides the strongest known globally controlled quantities. Hence, the greater strength of the nonlinear term relative to dissipation seen in the argument above is invariant under a large class of possible approaches.

The place where we butchered the argument is the first inequality in (35) where all of the three-dimensional geometry of the flow was thrown out. This geometry, or equivalently the geometry of vortex stretching, is arguably the key to progress. Yet, it remains poorly understood because it does not map easily into the language of continuity and compactness of mappings between topological vector spaces, and the latter forms the backbone of much of the theory of partial differential equations.

Appendix C. Spectral and Pseudo-Spectral Schemes

In this section, we briefly introduce spectral methods which are often the method of choice for the computational study of turbulence and blowup. Spectral methods rely on the Fourier series (or spectral decomposition) of the fluid fields.

Compared with alternative numerical schemes, spectral methods are fast, accurate, and easy to compare to many mathematical results. The last is because much of the mathematical analysis of partial differential equations uses spectral decompositions at some level. The main drawbacks of spectral methods are that they lose many of their advantages if used for anything except the simplest possible boundary geometries and, more serious in our case, they do not allow refinement if the most intense structures are very localized such as near developing singularities.

Under mild assumptions, the velocity field u has a unique representation in terms of the *Fourier series*

$$u(x,t) = \sum_{k \in \mathbb{Z}^d} u_k(t) e^{ik \cdot x} \tag{37}$$

where, for convenience, we have scaled our box-shaped periodic domain such that $\Omega = [0, 2\pi]^d$. Each of the *Fourier coefficients* u_k is a d-dimensional vector of complex numbers; the index k is referred to as the *wavenumber*. By taking the gradient of (37), we see that differentiation of u is equivalent to multiplication by ik on the Fourier side,

$$\nabla u(x, t) = \sum_{k \in \mathbb{Z}^d} i k \, u_k(t) \, e^{i k \cdot x}. \tag{38}$$

This observation can be turned into a numerical method by assuming that only the coefficients u_k with $|k| < n/2$ for some n are nonzero, so that the Fourier series involves no more than n^d summands. Consequently, the linear terms in (5) and (6) can be represented exactly by algebraic operations on this finite set of coefficients. The first apparent drawback is that a direct evaluation of the nonlinearity in the Fourier representation requires n^{2d} operations compared to n^d for the other terms, which would be prohibitively expensive.

Another problem stemming from the nonlinear term is that upon each new nonlinear evaluation, required for time advancement, the number of nonzero coefficients expands by a factor of 2^d. If these terms become large and cannot be neglected, then the required amount of computer memory will grow exponentially. Physically, this is perfectly reasonable, as it corresponds to the emergence of smaller scale structures as the flow evolves. This is called a *cascade* in the theory of turbulence. Cascades are naturally described in Fourier space, but are difficult to identify in the physical domain.

The inefficiency of calculating the nonlinearity in Fourier space is addressed by using the linear one-to-one correspondence between our set of n^d nonzero Fourier coefficients and the n^d values on equidistant mesh points in the physical domain. Since multiplication is cheap on the physical space mesh, we compute it there. Operations involving derivatives can be done efficiently in Fourier space. And the map between the Fourier representation and the physical space representation can be computed efficiently by the *fast Fourier transform*, or FFT, in just $n^d \ln n$ operations, i.e., we can map back and forth as needed without significant slowdown. Methods that split the operations in this way are known as *pseudo-spectral* codes.

Once the manner of calculating the spatial derivatives and the nonlinear interaction has been established, we have reduced the problem to solving a system of coupled ordinary differential equations. Combinations of well-known algorithms for the numerical solution of ordinary differential equations are then used to propagate the solution forward in time.

Pseudo-spectral approximations must be *dealiased* by setting to zero an appropriate set of the high wavenumber Fourier modes — the details are outside the scope of this discussion — to ensure that the results are mathematically equivalent to a truncation of the Fourier series as indicated earlier. Doing so ensures that quadratic invariants such as the energy in the Euler equations remain constants of the motion. However, errors still appear at the scale of the mesh spacing, that is in the high wavenumber Fourier coefficients, and in non-quadratic conservation laws such as the circulation — properties which can be monitored to assess the accuracy of a calculation [3, 11, 15].

Ultimately, the only way to ensure accuracy is to apply more resources, that is redo the calculations on finer meshes. In practice, when performing simulations at the limit of available resolution, a clear understanding of the biases of the chosen numerical scheme is as important as an understanding of the properties of the underlying partial differential equation. And often, the mathematical study of the numerical scheme is an interesting and worthwhile undertaking in its own right.

References

[1] J. Thomas Beale, Tosio Kato, and Andrew J. Majda, Remarks on the breakdown of smooth solutions for the 3-D Euler equations. *Communications in Mathematical Physics* **94**, 61–66 (1984)

[2] Claude Bardos and Edriss S. Titi, Euler equations for incompressible ideal fluids. *Russian Mathematical Surveys* **62**, 409–451 (2007)

[3] Miguel D. Bustamante and Robert M. Kerr, 3D Euler about a 2D symmetry plane. *Physica D: Nonlinear Phenomena* **237**, 1912–1920 (2008)

[4] Marco Cannone and Susan Friedlander, Navier: blow-up and collapse. *Notices of the American Mathematical Society* **50**, 7–13 (2003)

[5] Peter Constantin, On the Euler equations of incompressible fluids. *Bulletin of the American Mathematical Society* **44**, 603–621 (2007)

[6] Charles R. Doering, The 3D Navier–Stokes problem. *Annual Review of Fluid Mechanics* **41**, 109–128 (2009)

[7] Gregory L. Eyink, Dissipative anomalies in singular Euler flows. *Physica D: Nonlinear Phenomena* **237**, 1956–1968 (2008)

[8] Charles L. Fefferman, Existence & smoothness of the Navier–Stokes equation. Clay Mathematics Institute. http://www.claymath.org/millennium/Navier-Stokes_Equations/navierstokes.pdf (2000)

[9] Ciprian Foias, Oscar P. Manley, Ricardo M. S. Rosa, and Roger M. Temam, *Navier–Stokes Equations and Turbulence*. Cambridge University Press, Cambridge (2001)

[10] John D. Gibbon, The three-dimensional Euler equations: Where do we stand? *Physica D: Nonlinear Phenomena* **237**, 1894–1904 (2008)

[11] Tobias Grafke, Holger Homann, Jürgen Dreher, and Rainer Grauer, Numerical simulations of possible finite time singularities in the incompressible Euler equations: comparison of numerical methods. *Physica D: Nonlinear Phenomena* **237**, 1932–1936 (2008)

[12] John G. Heywood, Remarks on the possible global regularity of solutions of the three-dimensional Navier–Stokes equations. In: Giovanni P. Galdi, Josef Málek, and Jindřich Nečas (editors), *Progress in Theoretical and Computational Fluid Mechanics*, Paseky 1993, Pitman Research Notes in Mathematics Series, volume 308, pp. 1–32. Pitman, London (1994)

[13] Thomas Y. Hou and Ruo Li, Dynamic depletion of vortex stretching and non-blowup of the 3-D incompressible Euler equations. *Journal of Nonlinear Science* **16**, 639–664 (2006)

[14] Robert M. Kerr, Evidence for a singularity of the three-dimensional, incompressible Euler equations. *Physics of Fluids A* **5**, 1725–1746 (1993)
[15] Robert M. Kerr, *Computational Euler history*. Preprint. http://arxiv.org/abs/physics/0607148v2, 20 pages (July 19, 2006)
[16] Andrew J. Majda and Andrea L. Bertozzi, *Vorticity and Incompressible Flow*. Cambridge University Press, Cambridge (2002)
[17] Terence Tao, Why global regularity for Navier–Stokes is hard. http://terrytao.wordpress.com/2007/03/18/why-global-regularity-for-navier-stokes-is-hard/ (2007)
[18] Lloyd N. Trefethen, Ten digit problems. In: Dierk Schleicher and Malte Lackmann (editors), *An Invitation to Mathematics: From Competitions to Research*, pp. 119–136. Springer, Heidelberg (2011)

About the Hardy Inequality

Nader Masmoudi

Abstract. The Hardy inequality has a long history and many variants. Together with the Sobolev inequalities, it is one of the most frequently used inequalities in analysis. In this note, we present some aspects of its history, as well as some of its extensions and applications. This is a very active research direction.

1 Inequalities

Inequalities are among the main tools used in mathematics, and they can have very different roles within mathematics. They range from very classical inequalities (used in all fields of mathematics) such as the Cauchy-Schwarz inequality or the inequality between the arithmetic and the geometric mean to more specific ones. Inequalities can be important in their own right — as is often the case for instance in IMO problems — and they can be candidates for an "Oscar" for the best supporting actor within some other mathematical field. Indeed, in research an inequality is most often not the goal in itself but rather a tool to prove a theorem. Of course, in the olympiads, the same happens: one often has to use a known inequality to solve a problem. Sometimes, however, the problem is to prove a new inequality, so that the inequality is the goal itself and one has to be well equipped with proof methods of inequalities.

Of course, it happens that an inequality that is required somewhere in mathematics starts to take on a life of its own; and conversely, an inequality that has been investigated in its own interest may become useful somewhere else, perhaps unexpectedly. The Hardy inequality is an interesting such example, as we will see. It was discovered in an attempt to simplify the proof of

Nader Masmoudi
Courant Institute, New York University, 251 Mercer St, New York NY 10012, USA.
e-mail: masmoudi@cims.nyu.edu

another inequality, it was then studied in its own right and acquired several useful variants, and it eventually turned out to be extremely useful in the theory of partial differential equations.

Most inequalities have three forms: finite, infinite and integral. For instance, the *Hölder inequality* (which is a generalization of Cauchy-Schwarz) has these three different forms:

$$\sum_{i=1}^{n} a_i b_i \leq \left(\sum_{i=1}^{n} a_i^p\right)^{1/p} \left(\sum_{i=1}^{n} b_i^{p'}\right)^{1/p'}$$

$$\sum_{i=1}^{\infty} a_i b_i \leq \left(\sum_{i=1}^{\infty} a_i^p\right)^{1/p} \left(\sum_{i=1}^{\infty} b_i^{p'}\right)^{1/p'} \qquad (1)$$

$$\int_a^b f(x)g(x)\,dx \leq \left(\int_a^b f(x)^p\,dx\right)^{1/p} \left(\int_a^b g(x)^{p'}\,dx\right)^{1/p'}$$

for $\frac{1}{p} + \frac{1}{p'} = 1$. All the integrals here and in the sequel are taken in the sense of Lebesgue. The reader who is not familiar with this notion can assume that f and g are continuous or piecewise continuous functions defined on (a,b). Those who are not familiar with integrals at all can just focus on the versions involving series.

Even if not specified, all functions and series are supposed to be real-valued and non-negative in the whole article. Let us also recall the meaning of (1) (as well as all the inequalities that will be given later). The meaning of (1) is that if the right hand side is finite then the left hand side is also finite and the inequality holds. If the right hand side is infinite then the inequality does not say anything. So one can always assume that the right hand side is finite. Also, throughout the whole article, p will denote a real number with $1 < p < \infty$, and $p' > 1$ will denote the positive real number with $\frac{1}{p} + \frac{1}{p'} = 1$.

Some names may refer to different inequalities. In many cases, there is a relation between those inequalities. For instance, the name *Minkowski inequality* (in the integral form and for $p > 1$) usually refers to

$$\left[\int_a^b (f+g)^p\,dx\right]^{1/p} \leq \left[\int_a^b f^p\,dx\right]^{1/p} + \left[\int_a^b g^p\,dx\right]^{1/p} \qquad (2)$$

with equality if and only if f and g are proportional. One can also give the name Minkowski to the following double integral form

$$\left[\int_{I_2} \left(\int_{I_1} F(x_1,x_2)\,dx_1\right)^p dx_2\right]^{1/p} \leq \int_{I_1} \left(\int_{I_2} F(x_1,x_2)^p\,dx_2\right)^{1/p} dx_1 \qquad (3)$$

for any two intervals I_1 and I_2. The integrals in (2) occur frequently in analysis under the name of *p-norms* $\|f\|_p := \left[\int_a^b f^p\, dx\right]^{1/p}$, and (2) is simply the triangle inequality for this norm: $\|f+g\|_p \leq \|f\|_p + \|g\|_p$. The spaces of functions with finite *p*-norms are called L^p-spaces.

The two inequalities (2) and (3) can be interpreted as particular cases of a more general statement on measure spaces S_1 and S_2, namely the fact that $L_{x_2}^p(L_{x_1}^1) \supset L_{x_1}^1(L_{x_2}^p)$ for $p \geq 1$. One can recover (2) by taking S_1 to be reduced to two points and taking the counting measure on it.

Some inequalities may be reversed when we go from the infinite form to the integral one. For instance:

$$\left(\sum_{i=1}^\infty a_i^p\right)^{1/p} \leq \sum_{i=1}^\infty a_i ,$$

while

$$\int_a^b f(x)\, dx \leq (b-a)^{1-\frac{1}{p}} \left(\int_a^b f(x)^p\, dx\right)^{1/p}$$

where (a,b) is a finite interval. For readers familiar with Lebesgue spaces, these are just the inclusions $\ell^1 \subset \ell^p$ for sequence spaces and $L^p(a,b) \subset L^1(a,b)$ for function spaces. Of course these two inequalities can be seen as extensions of different sides of the following two inequalities for finite sums:

$$\left(\sum_{i=1}^N a_i^p\right)^{1/p} \leq \sum_{i=1}^N a_i \leq N^{\frac{p-1}{p}} \left(\sum_{i=1}^N a_i^p\right)^{1/p} .$$

Inequalities can come in the form of a strict inequality $<$ or of a non-strict inequality $\leq$. When a constant is involved, one of the important questions is to find the best constant and to study the case of equality. For instance in the case of the Hölder inequality (1) and dealing only with the positive case, this happens if $a_i^p = \lambda b_i^{p'}$ for all $i \in \mathbb{N}$ ($f(x)^p = \lambda g(x)^{p'}$ for all $x \in (a,b)$ in the integral case) for some fixed non-negative number λ. In particular, we note for later use that for any function g such that $\int_a^b g(x)^{p'}\, dx < \infty$, we have

$$\sup_f \int_a^b f(x)g(x)\, dx = \left(\int_a^b g(x)^{p'}\, dx\right)^{1/p'} \tag{4}$$

where the sup is taken over all f such that $\int_a^b f(x)^p\, dx = 1$. In (4) the sup is achieved at $f(x) = \dfrac{g(x)^{p'/p}}{\left(\int_a^b g(x)^{p'} dx\right)^{1/p}}$.

The proof (or the proofs) of most classical inequalities come from the convexity properties of some function such as $x^p, \exp(x), \ldots$ or from integration by parts (or summation by parts for series) or from the study of the maximum or minimum of some function (as in the proof of Theorem 2 below).

The aim of this note is to study one of these inequalities, namely the Hardy inequality (see [4, Chapter 9] and [2, 3] for earlier versions):

Theorem 1 (The Hardy Inequality).
1) If $A_n = a_1 + a_2 + \ldots + a_n$, then

$$\sum_{n=1}^{\infty} \left(\frac{A_n}{n}\right)^p < \left(\frac{p}{p-1}\right)^p \sum_{n=1}^{\infty} a_n^p \qquad (5)$$

unless all the a_n are zero. The constant is the best possible.
2) If $F(x) = \int_0^x f(t)\, dt$, then

$$\int_0^{\infty} \left(\frac{F(x)}{x}\right)^p dx < \left(\frac{p}{p-1}\right)^p \int_0^{\infty} f(x)^p\, dx \qquad (6)$$

unless $f \equiv 0$. The constant is the best possible.

As usual in Lebesgue theory, $f \equiv 0$ doesn't mean that $f = 0$ everywhere, but only on the complement of a set of (Lebesgue) measure zero. Of course for continuous functions, this makes no difference.

Notice here the similarity between the two inequalities (5) and (6). Indeed, $\frac{A_n}{n}$ is the arithmetic average of the sequence a up to the index n (this is often called the Césaro mean of the sequence (a_n), a frequent concept in summability theory), and $\frac{F(x)}{x}$ is the average of f over the interval $(0, x)$. Also, as stated after (1), the inequalities (5) and (6) mean that if the right hand side is finite then the left hand side is also finite and the inequality holds.

2 History of the Hardy Inequality

The first motivation of Hardy [2] was to find a simpler proof for the Hilbert inequality (see below). As stated in [4], Theorem 1 was discovered in the course of attempts to simplify the proofs of Hilbert's theorem that were known at the time. As a footnote, one can read "It was a considerable time before any really simple proof of Hilbert's double series theorem was found". Let us state without proof the Hilbert inequality:

Theorem 2 (The Hilbert Inequality).
1) If $\sum a_m^p \leq A$, $\sum b_n^{p'} \leq B$, the summation running from 1 to ∞, then

$$\sum_{m,n=1}^{\infty} \frac{a_m b_n}{m+n} < \frac{\pi}{\sin(\pi/p)} A^{1/p} B^{1/p'} \tag{7}$$

unless all the a_m are zero or all the b_n are zero. The constant is the best possible.

2) If $\int_0^\infty f(t)^p \, dt \leq A$, $\int_0^\infty g(t)^{p'} \, dt \leq B$, then

$$\int_0^\infty \int_0^\infty \frac{f(x)g(y)}{x+y} \, dx \, dy < \frac{\pi}{\sin(\pi/p)} A^{1/p} B^{1/p'} \tag{8}$$

unless $f \equiv 0$ or $g \equiv 0$. The constant is the best possible.

The determination of the best constant and the integral analogue are due to Schur. We only give an elementary proof of (7) in the case $p = 2$ and $a = b$ (here and elsewhere, we simply write a or b for finite or infinite sequences (a_n) and (b_n), so $a = b$ means that $a_n = b_n$ for all n). Our proof is based on the theory of maxima and minima of functions of several variables (see [4, Appendix III]; for a full proof, see Chapter 9). We will prove a slightly stronger version of (7), namely that

$$\sum_{m,n=0}^{\infty} \frac{a_m a_n}{m+n+1} \leq \pi \sum_{n=0}^{\infty} a_n^2 . \tag{9}$$

We may suppose that more than one a_n is different from 0, otherwise the inequality is trivial. Consider the two functions

$$F(a) = \sum_{m,n=0}^{N} \frac{a_m a_n}{m+n+1}, \qquad G(a) = \sum_{n=0}^{N} a_n^2$$

defined for finite sequences $a = (a_0, a_1, \ldots, a_N) \in [0, +\infty)^{N+1}$. We want to prove that $F(a) < \pi G(a)$ for any $a \neq 0$. For each $t > 0$, we maximize the function F on the set of all sequences a with $G(a) = t$. This set is clearly compact and hence F attains its maximum $F^* = F^*(t)$ at some point a.

We would like to deduce from that a Lagrange equation. Hence, we first need to show that all the a_n are positive, so that a is away from the boundary of its domain. Indeed, if any $a_n = 0$, then a small increment δ in a_n produces an increase of δ^2 in G and one of order δ in F. Hence, denoting $b = \sqrt{\frac{t}{t+\delta^2}}(a + \delta e_n)$, where $e_n = (0, \ldots, 1, 0, \ldots)$ with 1 at the n-th position, we see that $G(b) = t$ and $F(b) > F(a)$.

Hence, necessarily all the a_n are positive and we deduce from the maximality of $F(a)$ that there is a Lagrange[1] multiplier λ satisfying

$$\frac{\partial F}{\partial a_n} - \lambda \frac{\partial G}{\partial a_n} = 0 \tag{10}$$

for all $n \leq N$. Hence, for $n \leq N$, we have

$$\sum_{m=0}^{N} \frac{a_m}{m+n+1} = \lambda a_n . \tag{11}$$

Multiplying by a_n and adding, we get $F(a) = \lambda t$.

Let $a_m \sqrt{m + \frac{1}{2}}$ be maximal for $m = m_0$. Then, for $n = m_0$ in (11), we get

$$\lambda a_{m_0} = \sum_{m=0}^{N} \frac{a_m}{m + m_0 + 1} \leq a_{m_0} \sqrt{m_0 + \frac{1}{2}} \sum_{m=0}^{N} \frac{1}{(m + m_0 + 1)\sqrt{m + \frac{1}{2}}}$$

$$\leq a_{m_0} \sqrt{m_0 + \frac{1}{2}} \int_{-1/2}^{N+1/2} \frac{dx}{(x + m_0 + 1)\sqrt{x + \frac{1}{2}}}$$

$$= a_{m_0} \int_{0}^{\sqrt{\frac{N+1}{m_0 + \frac{1}{2}}}} \frac{2\, dy}{y^2 + 1} < a_{m_0} \int_{0}^{\infty} \frac{2\, dy}{y^2 + 1} = \pi a_{m_0} .$$

Here we used in the second line that $\left((x + m_0 + 1)\sqrt{x + \frac{1}{2}}\right)^{-1}$ is convex in x; from the second to the third line we used the substitution $y = \sqrt{(x + \frac{1}{2})/(m_0 + \frac{1}{2})}$, and for the last integral we used a well known formula (which follows by substituting $y = \tan z$).

From the inequality above, we deduce that $\lambda < \pi$, and thus $F(a) < \pi G(a)$ holds for any $a \neq 0$. Sending N to infinity, we deduce the infinite series version, namely the fact that (9) holds. This also yields (7) (in our special case) by replacing a_n by a_{n-1} and using that $\frac{1}{m+n} < \frac{1}{m+n-1}$. □

---

[1] The idea of Lagrange multipliers is simple: at every point a, the vector of partial derivatives $\left(\frac{\partial F}{\partial a_0}, \frac{\partial F}{\partial a_1}, \ldots, \frac{\partial F}{\partial a_N}\right)$, called the *gradient of F at a*, points in the direction (in the space of sequences a) of maximal increase of F. The same is true for G. Since we only allow sequences a for which G assumes the constant value t, the values of a are restricted to an N-dimensional hypersurface (think of a 2-dimensional surface in ordinary 3-space). The gradient of G is perpendicular to this hypersurface (at every point, the direction of maximal increase of G is perpendicular to the hypersurface of constant values). If the gradient of F was not perpendicular to this hypersurface, then there would be a direction along the hypersurface along which F could increase, and this cannot happen at a point a where F is maximal. Hence the gradients of F and G must both be perpendicular to the hypersurface and thus be parallel, up to sign, and the existence of λ follows. — Those who are not familiar with partial derivatives like $\partial F / \partial a_n$ or with the gradient may wish to consult Appendix A.1 in the contribution [6] in this volume.

3 Proof of the Hardy Inequality

Here, we give a proof of Theorem 1. In the series case, the proof is due to Elliott. Let us write $\alpha_n = A_n/n$ and $\alpha_0 = 0$. We have

$$\alpha_n^p - \frac{p}{p-1} a_n \alpha_n^{p-1} = \alpha_n^p - \frac{p}{p-1}\left[n\alpha_n - (n-1)\alpha_{n-1}\right]\alpha_n^{p-1}$$

$$= \alpha_n^p\left(1 - \frac{np}{p-1}\right) + \frac{(n-1)p}{p-1}\alpha_n^{p-1}\alpha_{n-1}$$

$$\leq \alpha_n^p\left(1 - \frac{np}{p-1}\right) + \frac{n-1}{p-1}\left[(p-1)\alpha_n^p + \alpha_{n-1}^p\right] \quad (12)$$

$$= \frac{1}{p-1}\left[(n-1)\alpha_{n-1}^p - n\alpha_n^p\right]$$

where we have used the *Young inequality* on the third line, namely $xy \leq \frac{x^p}{p} + \frac{y^{p'}}{p'}$ with $y = \alpha_n^{p-1}$ and $x = \alpha_{n-1}$.

Summing from 1 to N yields a telescoping sum on the right hand side, so we get

$$\sum_{n=1}^{N} \alpha_n^p - \frac{p}{p-1}\sum_{n=1}^{N}\alpha_n^{p-1}a_n \leq -\frac{N\alpha_N^p}{p-1} \leq 0 \quad (13)$$

and hence, by the Hölder inequality, we get

$$\sum_{n=1}^{N}\alpha_n^p \leq \frac{p}{p-1}\sum_{n=1}^{N}\alpha_n^{p-1}a_n \leq \frac{p}{p-1}\left(\sum_{n=1}^{N}a_n^p\right)^{1/p}\left(\sum_{n=1}^{N}\alpha_n^p\right)^{1/p'}. \quad (14)$$

Dividing by the last factor (which is positive, otherwise there is nothing to prove) and raising to the power p yields the result in the finite case. In particular, we see that $\sum_{n=1}^{\infty}\alpha_n^p$ is finite if $\sum_{n=1}^{\infty}a_n^p$ is finite. Replacing N by ∞ in (13) and (14) yields

$$\sum_{n=1}^{\infty}\alpha_n^p \leq \frac{p}{p-1}\left(\sum_{n=1}^{\infty}a_n^p\right)^{1/p}\left(\sum_{n=1}^{\infty}\alpha_n^p\right)^{1/p'} \quad (15)$$

and the inequality is strict unless a_n^p and α_n^p are proportional, i.e unless $a_n = C\alpha_n$ where C is independent of n. Without loss of generality, we can assume that $a_1 \neq 0$. Otherwise, we can replace a_{n+1} by a_n and the inequality becomes weaker. Hence, $C = 1$ and we infer that $A_n = na_n$ which is only possible if all the a are equal which is inconsistent with the convergence of $\sum \alpha_n^p$. Hence, (5) holds.

To prove that the constant is optimal, we take $a_n = n^{-1/p}$ for $n \leq N$ and $a_n = 0$ for $n > N$, where N is some positive integer that will be specified below. Hence, $\sum_{n=1}^{N} a_n^p = \sum_{n=1}^{N} \frac{1}{n}$ and

$$A_n = \sum_{k=1}^{n} k^{-1/p} > \int_1^n x^{-1/p}\, dx = \frac{p}{p-1}\left[n^{\frac{p-1}{p}} - 1\right] \quad (n \leq N),$$

so

$$\left(\frac{A_n}{n}\right)^p > \left(\frac{p}{p-1}\right)^p \frac{1-\varepsilon_n}{n} \quad (n \leq N),$$

where ε_n only depends on n (and not on N) and ε_n goes to zero when n goes to infinity.

Now, let $\varepsilon > 0$ be given and let n_0 be some positive integer such that $\varepsilon_n < \varepsilon$ whenever $n \geq n_0$. Choose N such that $\sum_{n=1}^{N} \frac{1}{n} > \frac{1}{\varepsilon} \sum_{n=1}^{n_0-1} \frac{1}{n}$ (this is possible because the harmonic series diverges). Then for the sequence (a_n) as defined above, we have

$$\sum_{n=1}^{\infty} \left(\frac{A_n}{n}\right)^p > \sum_{n=n_0}^{N} \left(\frac{A_n}{n}\right)^p > \left(\frac{p}{p-1}\right)^p \sum_{n=n_0}^{N} \frac{1-\varepsilon_n}{n} > \left(\frac{p}{p-1}\right)^p \sum_{n=n_0}^{N} \frac{1-\varepsilon}{n}$$

$$= (1-\varepsilon)\left(\frac{p}{p-1}\right)^p \sum_{n=n_0}^{N} \frac{1}{n} > (1-\varepsilon)^2 \left(\frac{p}{p-1}\right)^p \sum_{n=1}^{N} \frac{1}{n}$$

$$= (1-\varepsilon)^2 \left(\frac{p}{p-1}\right)^p \sum_{n=1}^{\infty} a_n^p.$$

If we let ε tend to 0, this shows that the constant $\left(\frac{p}{p-1}\right)^p$ is optimal. An alternative choice is to take $a_n = n^{-1/p-\varepsilon}$ for all n and to send ε to zero.

Now, we turn to the proof of the integral inequality. Integrating by parts, we have

$$\int_0^X \left(\frac{F(x)}{x}\right)^p dx = -\frac{1}{p-1} \int_0^X F(x)^p \frac{d}{dx}(x^{1-p})\, dx$$

$$= \left[-\frac{x^{1-p}F(x)^p}{p-1}\right]_0^X + \frac{p}{p-1} \int_0^X \left(\frac{F(x)}{x}\right)^{p-1} f(x)\, dx$$

$$\leq \frac{p}{p-1} \int_0^X \left(\frac{F(x)}{x}\right)^{p-1} f(x)\, dx$$

since the integrated term (first term on the second line) vanishes at $x = 0$ in virtue of $F(x) = O(x)$ if we assume that f is continuous on $[0, \infty)$. If we only assume that f^p is integrable, we can get the same conclusion using that

$F(x) \le (\int_0^x f(t)^p \, dt)^{1/p} x^{\frac{p-1}{p}}$ by the Hölder inequality and that $\int_0^x f(t)^p \, dt$ goes to zero when x goes to 0.

Sending X to infinity and using the Hölder inequality, we get as in the series case that the strict inequality (6) holds unless $x^{-p}F^p$ and f^p are proportional, which is impossible, since it would make f a power of x and hence $\int f^p \, dx$ divergent. The proof that the constant is optimal can be done by looking at $f_\varepsilon(x) = 0$ for $x < 1$ and $f_\varepsilon(x) = x^{-1/p-\varepsilon}$ for $x \ge 1$ and then sending ε to zero. Two other choices consist in taking $g_\varepsilon(x) = 0$ for $x \ge 1$ and $g_\varepsilon(x) = x^{-1/p+\varepsilon}$ for $x < 1$ or $h_\varepsilon(x) = x^{-1/p}$ for $x \in (\varepsilon, \frac{1}{\varepsilon})$ and $h_\varepsilon(x) = 0$ elsewhere and then sending ε to zero. □

Remark 1. Notice that (12) is a sort of integration by parts similar to the one used in the Abel transformation.

Remark 2. In the limit $p \searrow 1$, the inequalities in Theorem 1 are void because both sides are infinite, unless a or f are identically zero. Indeed, if $a_k > 0$, then $A_n \ge a_k$ for $n \ge k$, and we have a diverging harmonic series as lower bound. On the right hand side, clearly $p/(p-1) \to \infty$.

4 Variants of the Hardy Inequality

The Case of a Decreasing Function. If in Theorem 1, we assume that f is non-increasing, then we get the following two-sided inequality

$$\left(\frac{p}{p-1}\right) \int_0^\infty f(x)^p \, dx \le \int_0^\infty \left(\frac{F(x)}{x}\right)^p \, dx < \left(\frac{p}{p-1}\right)^p \int_0^\infty f(x)^p \, dx \, . \tag{16}$$

To prove the left inequality, we notice that

$$\frac{d}{dt}[F(t)^p] = pf(t)F(t)^{p-1} \ge pf(t)^p t^{p-1}$$

where we have used that f is non-increasing. Integrating between 0 and x, we get that

$$F(x)^p \ge p \int_0^x f(t)^p t^{p-1} \, dt \, .$$

Hence,

$$\int_0^\infty \left(\frac{F(x)}{x}\right)^p dx \geq p \int_0^\infty x^{-p} \int_0^x f(t)^p t^{p-1} dt\, dx$$

$$= p \int_0^\infty \left(\int_t^\infty x^{-p} dx\right) f(t)^p t^{p-1} dt$$

$$= \frac{p}{p-1} \int_0^\infty f(t)^p dt\ .$$

The Weighted Hardy Inequality.

Theorem 3 (The Weighted Hardy Inequality).
If $p > 1$ and $r \neq 1$ and $F(x)$ is defined by

$$F(x) = \int_0^x f(t)\, dt \quad \text{if} \quad r > 1, \qquad F(x) = \int_x^\infty f(t)\, dt \quad \text{if} \quad r < 1,$$

then

$$\int_0^\infty x^{-r} F(x)^p\, dx \leq \left(\frac{p}{|r-1|}\right)^p \int_0^\infty f(x)^p x^{p-r}\, dx \qquad (17)$$

and the constant is the best possible.

Again, here (17) means that if the right hand side is finite, then the left hand side is also finite and the inequality holds.

We only give the proof in the case $r > 1$. The proof in the second case is very similar. The proof uses the Minkowski inequality (3):

$$\left(\int_0^\infty x^{-r}\left(\int_0^x f(t)\, dt\right)^p dx\right)^{1/p} = \left(\int_0^\infty x^{p-r}\left(\int_0^1 f(sx)\, ds\right)^p dx\right)^{1/p}$$

$$\leq \int_0^1 \left(\int_0^\infty f(sx)^p x^{p-r}\, dx\right)^{1/p} ds$$

$$= \int_0^1 s^{-\frac{1+p-r}{p}} \left(\int_0^\infty f(y)^p y^{p-r}\, dy\right)^{1/p} ds$$

$$= \frac{p}{r-1}\left(\int_0^\infty f(y)^p y^{p-r}\, dy\right)^{1/p}.$$

Here, we have made the change of variables $t = sx$ in the first line and $y = sx$ in the third one. This yields (17). Notice also that this gives another proof of the original Hardy inequality when $r = p$. We leave it to the reader to check that the constant is optimal. We also point out that if $p = 1$, then by a simple integration by parts, one can see that (17) is actually an equality. □

About the Hardy Inequality

The next theorem is a generalization of (17) in the sense that it gives a necessary and sufficient condition on the non-negative functions u and v so that the weighted Hardy inequality (18) holds [8]. Again u and v are supposed to be non-negative measurable functions on the interval $(0, b)$. The reader who is not familiar with this notion can suppose them to be continuous on the open interval $(0, b)$.

Theorem 4 (The Generalized Weighted Hardy Inequality).
Let $p > 1$ and $0 < b \leq \infty$. The inequality

$$\int_0^b \left(\int_0^x f(t)\, dt \right)^p u(x)\, dx \leq C \int_0^b f(x)^p v(x)\, dx \tag{18}$$

holds for any measurable (or just continuous) function $f(x) \geq 0$ on $(0, b)$ if and only if

$$A = \sup_{r \in (0,b)} \left(\int_r^b u(x)\, dx \right)^{1/p} \left(\int_0^r v(x)^{1-p'}\, dx \right)^{1/p'} < \infty. \tag{19}$$

Moreover, the best constant C in (18) satisfies $A \leq C^{1/p} \leq p^{1/p}(p')^{1/p'} A$.

We only prove that the condition is necessary and leave the other part to the reader.

We assume that (18) holds for any f such that $\int_0^b f(x)^p v(x)\, dx < \infty$. For any such f, when applying (18) to $f_r = f\chi_{(0,r)}$, we get

$$\int_r^b u(x)\, dx \left(\int_0^r f(t)\, dt \right)^p \leq C \int_0^r f(x)^p v(x)\, dx. \tag{20}$$

Let us denote

$$M = \sup_f \int_0^r f(t)\, dt \tag{21}$$

where the sup is taken over all functions f such that $\int_0^r f(x)^p v(x)\, dx = 1$. Using (4), we deduce easily that

$$M = \left(\int_0^r v^{1-p'}\, dt \right)^{1/p'}.$$

Indeed, introducing $h(t) = f(t)v(t)^{1/p}$, we see that

$$M = \sup_h \int_0^r h(t)v(t)^{-1/p}\, dt$$

where the sup is taken over all h such that $\int_0^r h(t)^p\, dt = 1$. The reader familiar with weighted L^p space can deduce this directly.

Hence, we deduce from (20) that

$$\int_r^b u(x)\,dx \left(\int_0^r v^{1-p'}\,dt\right)^{p/p'} \leq C. \qquad (22)$$

Therefore, we see that (19) holds and that $A^p \leq C$.

The proof that the condition (19) is sufficient and that the optimal constant C satisfies $C^{1/p} \leq p^{1/p}(p')^{1/p'} A$ is left to the reader. □

The Limit p Going to 1 and to ∞. As stated in Remark 2, the Hardy inequality (6) does not hold for $p = 1$. However, (18) can be extended to $p = 1$, if we replace the second term on the right hand side of (19) by $\sup_{x\in(0,r)} \frac{1}{v}$. We also leave this to the reader.

Next, we would like to consider the limit p going to infinity in (5) and (6). We have the following theorem:

Theorem 5 (The Hardy Inequality for $p = \infty$).
1) If $a_n \geq 0$, then

$$\sum_{n=1}^{\infty} (a_1 a_2 \ldots a_n)^{1/n} < e \sum_{n=1}^{\infty} a_n \qquad (23)$$

unless all the a are zero. The constant is the best possible.

2) If $f(x) \geq 0$, then

$$\int_0^{\infty} \exp\left(\frac{1}{x}\int_0^x \log f(t)\,dt\right)\,dx < e\int_0^{\infty} f(x)\,dx \qquad (24)$$

unless $f \equiv 0$. The constant is the best possible.

We can prove (23) with a $\leq$ instead of $<$ by a passage to the limit. Replacing a_n^p by a_n in (5), we get

$$\sum_{n=1}^{\infty} \left(\frac{a_1^{1/p} + a_2^{1/p} + \ldots + a_n^{1/p}}{n}\right)^p < \left(\frac{p}{p-1}\right)^p \sum_{n=1}^{\infty} a_n. \qquad (25)$$

Sending p to infinity, we deduce that (23) holds with a $\leq$. To prove that it is actually $<$ unless all the a_n are zero requires a different proof [4].

For the proof of (24), we use the same idea, namely we replace f^p by f in (6) and then send p to infinity. We just have to observe that

$$\lim_{p\to\infty} \left(\frac{1}{x}\int_0^x f(t)^{1/p}\,dt\right)^p = \exp\left(\frac{1}{x}\int_0^x \log f(t)\,dt\right). \qquad (26)$$

which follows for example by l'Hôpital's rule. □

5 Applications

One branch of mathematics where inequalities, including the Hardy inequality in its various forms, are essential tools is the theory of partial differential equations (PDEs). Partial differential equations are equations which contain derivatives of an unknown function in more than one variable; many problems in mathematics, physics, engineering, and similar fields give rise to PDEs. A fundamental mathematical question is whether a PDE has solutions and, if so, if these are unique. The motor for essentially all such theorems are inequalities, and it is here that the Hardy inequality is often put to good use.

Many important PDEs describe the time development (evolution) of interesting quantities. In such cases, a second question is whether solutions to these equations exist forever, or whether they "blow up" in finite time, so that some quantities become infinite and solutions fail to exist thereafter. A simple case of blow-up in finite time has been described by Nick Trefethen in this volume [10, Section 4 "Blow-Up"]. The importance of the blow-up question in fluid dynamics is discussed in the contribution by Bob Kerr and Marcel Oliver [6], especially Sections 4 and 5; their Appendix B demonstrates explicitly how inequalities come into play.

We would like to present here a few specific applications of where the Hardy inequalities come up in the theory of partial differential equations. This section is meant to give you a rough idea; readers who are not familiar with some notation or background are encouraged to read it like "science fiction".

Boundary Traces of Non-Smooth Functions. In the theory of partial differential equations it is often useful to generalize the class of functions from which to seek a solution beyond functions for which all partial derivatives which appear in the PDE exist and are continuous. In this case, the equation must be interpreted in a carefully defined averaged or "weak" sense. One class of functions $f: \Omega \to \mathbb{R}$ which is frequently encountered is the Lebesgue space $L^p(\Omega)$ (recall that L^p- functions are defined only "almost everywhere"), another is the Sobolev space

$$W^{1,p}(\Omega) := \{f \in L^p(\Omega) \,|\, \nabla f \in L^p(\Omega)\} \,,$$

where ∇f is the gradient of f (see [6, Appendix A.1] in this volume); the superscript 1 in $W^{1,p}$ says that we want to control the function f and its first derivatives on the domain Ω in the L^p sense (even though f may be undefined on a set of measure zero, it is still possible that almost everywhere they have derivatives in a certain sense, called "distributions"; we require that these derivatives must also be L^p-functions).

When the system described by the PDE is defined on a domain Ω with smooth boundary $\partial\Omega$, the PDE must usually be augmented by boundary conditions such as by specifying which values the solution must assume on the boundary. Hence, we need to establish a relationship between functions (from a certain space) defined on the entire domain and their restrictions to the boundary. This is done by means of a *trace operator* γ which, for smooth functions on Ω, is simply the restriction to the boundary. For functions $f \in L^p(\Omega)$, the trace is not well defined because the boundary is a set of Lebesgue measure zero, hence is not "seen" by the L^p norm. On the other hand, γ extends to a continuous operator from $W^{1,p}(\Omega)$ into a certain space that we shall not define here (a "Besov space of fractional order"). The proof of this result is based on the fact that $\nabla f \in L^p(\Omega)$ implies that

$$\frac{f(x) - \gamma f(\pi(x))}{d(x)} \in L^p(\Omega) , \qquad (27)$$

where $d(x)$ is the distance of x to the boundary of the domain and $\pi(x)$ denotes the boundary point of Ω nearest to x (defined for x close to the boundary). This can be deduced from the Hardy inequality written in the perpendicular direction to the boundary.

Compressible Euler Equations with Degenerate Density. When partial differential equations involve coefficients that go to zero near the boundary of Ω, we often need to control quantities like $f(x)/d(x)$ that diverge as x tends to the boundary. In all these cases, estimating $f(x)/d(x)$ in appropriate spaces of functions requires some control on the derivatives of f, and this is achieved using Hardy or Hardy–Sobolev estimates.

One such example arises when studying acoustic waves within a bubble of a compressible fluid surrounded by vacuum. Generally, the dynamics of a fluid without friction is described by the Euler equations. They are discussed, mainly in their incompressible form, in the contribution [6] by Bob Kerr and Marcel Oliver in this volume. The compressible Euler equations describe the motion of a gas whose density at time t and position $\boldsymbol{x}$ is denoted $\rho(\boldsymbol{x},t)$, its velocity by $\boldsymbol{u}(\boldsymbol{x},t)$, and the pressure by $p(\boldsymbol{x},t)$. Written in terms of these so-called Eulerian variables and omitting arguments, the equations read

$$\rho \frac{\partial \boldsymbol{u}}{\partial t} + \rho \boldsymbol{u} \cdot \operatorname{grad} \boldsymbol{u} + \operatorname{grad} p = 0 , \qquad (28\mathrm{a})$$

$$\frac{\partial \rho}{\partial t} + \operatorname{div}(\rho \boldsymbol{u}) = 0 . \qquad (28\mathrm{b})$$

The first equation is the same as (5a) of [6]. The second equation (28b) expresses conservation of mass: if more fluid flows into a neighborhood of a point than flows out, then the density increases. For an incompressible homogeneous flow where $\rho = \text{const}$, it reduces to $\operatorname{div} \boldsymbol{u} = 0$, cf. [6].

Compared to the incompressible Euler system, there is one more unknown function, the density ρ. Correspondingly, an extra condition is needed, which can be supplied, for example, by assuming the equation of state for an isentropic gas $p = K\rho^\gamma$ where $\gamma > 1$ is called the adiabatic exponent and K is some constant of proportionality.

When we consider a bubble of gas surrounded by vacuum, the density is zero outside the bubble — the equation "degenerates" toward the boundary of the bubble.

It is convenient to rewrite the system in terms of the *flow map* $\boldsymbol{\Phi}(\boldsymbol{\xi}, t)$ which describes the position of a fluid particle at time t when it started from position $\boldsymbol{\xi}$ at time $t = 0$. In this form (which we omit), turning "Eulerian variables" into "Lagrangian variables", the resulting equations look like a nonlinear wave equation. (The linear wave equation is one of the the best known PDEs. It describes, for example, the propagation of light and of acoustic waves in air under typical conditions.) In this form, the initial density turns into an external parameter in such a way that, under the assumption of a vacuum in the exterior of the bubble, it behaves like $d(x)$ near the boundary, hence is responsible for the degeneracy.

This degeneration is best explained in a linear toy model system: the equation in a single space variable $x \in (0, 1) \subset \mathbb{R}$

$$w^\alpha \frac{\partial^2 \Phi}{\partial t^2} + \frac{\partial}{\partial x}\left(w^{1+\alpha} \frac{\partial \Phi}{\partial x}\right) = 0 \tag{29}$$

where $w = x(1-x)$, $\alpha = 1/(\gamma - 1) > 0$, and $\Phi(x, t) \in \mathbb{R}$ is the (one-dimensional) flow map. This is a very simple equation, from the point of view of those dealing with partial differential equations. Note again that this equation degenerates when x tends to the boundary $\{0, 1\}$ (because then w tends to 0). The study of even this relatively simple partial differential equation requires the use of the weighted Hardy inequality; see [5].

There are many further examples for partial differential equations coming e.g. from physics that need the use of the Hardy inequality. These include equations describing thin films or porous media [1], or so-called "Fokker-Planck equations" for polymers [9].

6 Conclusion

In this article, we presented various aspects of the Hardy inequality and its generalizations. This inequality has a very long and interesting history. It is a prominent example of an inequality that was first studied in order to simplify the proof of some other inequality (Hilbert's inequality), before the relevance of this inequality for other areas of mathematics was realized and it assumed an important role, notably in partial differential equations. This is definitely a very active research direction where one tries to find new inequalities with precise applications in mind.

References

[1] Lorenzo Giacomelli, Hans Knüpfer, and Felix Otto, Smooth zero-contact-angle solutions to a thin-film equation around the steady state. *Journal of Differential Equations* **245**(6), 1454–1506 (2008)

[2] Godfrey H. Hardy, Notes on some points in the integral calculus. XLI. On the convergence of certain integrals and series. *Messenger of Mathematics* **45**, 163–166 (1915)

[3] Godfrey H. Hardy, Notes on some points in the integral calculus. LX. An inequality between integrals. *Messenger of Mathematics* **54**, 150–156 (1925)

[4] Godfrey H. Hardy, John E. Littlewood, and George Pólya, *Inequalities*. Cambridge Mathematical Library. Cambridge University Press, Cambridge (1988); reprint of the 1952 edition

[5] Juhi Jang and Nader Masmoudi, *Well-posedness of compressible Euler equations in a physical vacuum*. Preprint. http://arxiv.org/abs/1005.4441, 35 pages (May 24, 2010)

[6] Robert M. Kerr and Marcel Oliver, The ever-elusive blowup in the mathematical description of fluids. In: Dierk Schleicher and Malte Lackmann (editors), *An Invitation to Mathematics: From Competitions to Research*, pp. 137–164. Springer, Heidelberg (2011)

[7] Alois Kufner, Lech Maligranda, and Lars-Erik Persson, The prehistory of the Hardy inequality. *American Mathematical Monthly* **113**(8), 715–732 (2006)

[8] Alois Kufner, Lech Maligranda, and Lars-Erik Persson, *The Hardy inequality. About Its History and Some Related Results*. Vydavatelský Servis, Plzeň (2007)

[9] Nader Masmoudi, Well-posedness for the FENE dumbbell model of polymeric flows. *Communications on Pure and Applied Mathematics* **61**(12), 1685–1714 (2008)

[10] Lloyd N. Trefethen, Ten digit problems. In: Dierk Schleicher and Malte Lackmann (editors), *An Invitation to Mathematics: From Competitions to Research*, pp. 119–136. Springer, Heidelberg (2011)

The Lion and the Christian, and Other Pursuit and Evasion Games

Béla Bollobás

Abstract. In this note we shall show that a playful question in recreational mathematics can quickly lead to mathematical results and unsolved problems.

1 An Arena in Rome

Fig. 1. An aged Lion and an agile young Christian in the arena.

Béla Bollobás
Department of Pure Mathematics and Mathematical Statistics, University of Cambridge, Cambridge CB3 0WB, UK; and Department of Mathematical Sciences, University of Memphis, Memphis TN 38152, USA. e-mail: B.Bollobas@dpmms.cam.ac.uk

It was a beautiful spring day in the second year of the reign of Marcus Ulpius Nerva Traianus, Emperor of Rome. The gladiatorial games put on to celebrate the Emperor's victory over Dacebalus, King of Dacia, and the conquest of his entire kingdom had been in full swing for sixty days, and to the joy of the good citizens of Rome showed no sign of abating. The amazing Colosseum that had been built a few years before was almost full, and the people were eagerly awaiting the treat this beautiful new day would bring them.

It all started well: the gladiators were magnificent — they were skilful and extremely brave, and many a fighter received his freedom from the Emperor. However, when it came to one of the favourite parts of the show, *men against beasts*, there was disappointment in the air: only one Lion and one Christian appeared in the arena. The Lion was big enough, but on a closer examination he turned out to be old, way past his prime, while the Christian was a fit young man. It soon transpired that the Lion and the Christian had the same top speed, so the good citizens of Rome began to wonder whether the Lion would ever catch the Christian.

While watching this pathetic spectacle, the more mathematical-minded citizens of Rome came up with the following problem they were confident they could solve within a few minutes.

The Lion and Christian Problem. *A Lion and a Christian (each considered to be a single point) move about in a closed circular disc with the same maximum speed. Can the Lion catch the Christian in finite time?*

As we shall see, this problem is not as easy as it looks: it is a prime example of the large family of *pursuit and evasion* problems. In this brief paper we shall discuss some aspects of these problems.

Note that in order to turn their conundrum into a mathematical problem, the citizens have made some simplifications usual in mathematics and physics: the Lion and the Christian have been turned into *points*, and the Colosseum has become a *circular* closed disc, rather than an oval, which would be a better approximation of the shape of the arena. The disc is assumed to be *closed*, so that the points corresponding to the Lion and Christian are allowed to be on the circle bounding the disc. Rather importantly, the catch is required to be in *finite time*. (Needless to say, it is assumed that at the start of the game the Lion and the Christian are not in the same spot.) The question is about a 'clever' Lion and a 'clever' Christian: the Lion 'plays' as well as possible, and so does the Christian.

Contrary to the story above, this question was not invented by the good citizens of Rome close to two thousand years ago, but in the 1930s by the German–British mathematician Richard Rado, who called it the *Lion and Man* problem. Its solution (the second solution below) remained the standard answer for about twenty years, when the Russian–British mathematician Abram S. Besicovitch, the Rouse Ball Professor in Cambridge, found a brilliant and unexpected solution (the third solution below). After this, the question and its answer by 'Bessie', as Besicovitch was known affectionately, were extensively popularized by the great British mathematician John E. Littlewood, Bessie's predecessor in the Rouse Ball chair, in his *Miscellany* [6] (see also [2]).

In the next section we sketch three solutions to this problem; the section after that is about various extensions of it, the fourth section is about some finer points of pursuit and evasion games, and the last section is about further results and open problems.

2 Solutions

In this section we shall give three solutions to the LC problem, coming to different conclusions.

First Solution: Curve of Pursuit. Clearly, the best strategy for the Lion is to run right towards the Christian. What should the Christian do? He should run round the perimeter of the disc at full speed. What develops then is that the Christian runs around the circle, and the Lion is running along the 'curve of pursuit'. Although this curve is not too easy to describe explicitly, even when the Lion starts at the centre of the disc and the Christian starts his run on the boundary, as in Figure 2, one can show that during this chase the Lion gets arbitrarily close to the Christian, *without ever catching him*.

Conclusion. *The Christian wins the LC game.* □

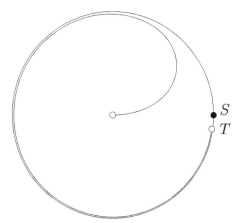

Fig. 2. The curve of pursuit when the Lion starts from the centre and the Christian runs a full circle starting and ending at S. When the Christian gets back to S, the Lion is not far away, at T.

Curve of pursuit problems, like the one above, have been around in mathematics for almost three centuries, with the pursuer and the pursuit having different speeds. The standard formulation of these *pure pursuit* problems is that of a dog racing towards his master who is walking in a field. (See Puckette [8] and Nahin [7].)

To give two other solutions, we shall need some notation. First, we shall write B for the position of the 'Beast', the Lion, and C for that of the Christian, suppressing the dependence of these points on the time. We may and shall assume that the action takes place in the unit disc D with centre O, and the maximum speed of the contestants is 1. Also, we shall take D to be *closed*, i.e. we shall take it together with its boundary. (In fact, much of the time it will make no difference whether D is closed or not, although occasionally, like in the first solution we have just seen, and in the second solution below, it is convenient to take the boundary circle as part of our disc D.)

Second Solution: Stay on the Radius. The Lion decides to be not as greedy as in the first solution, but adopts the cunning plan of *staying on the segment OC* and, subject to this constraint, running towards the Christian at maximum speed. What happens if, as before, the Christian races along the boundary of D at full speed? Let us assume, for simplicity, that the Christian, C, starts at a point S of the big (radius 1) circle and runs in anticlockwise direction, and the Lion, B, starts at its centre, O.

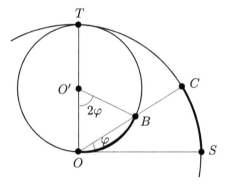

Fig. 3. The Lion's path when he starts from the centre and the Christian from S. When the Christian, running along the big circle, reaches T, so does the Lion.

Claim. If the Lion follows his 'stay-on-the-radius' strategy, then he runs along the small (radius $1/2$) circle touching the line OS at O and the big circle at T, where the ST-arc of the big circle is a quarter-circle, as in Figure 3. Even more, wherever C is on the arc ST of the big circle, B is the intersection of the segment OC with the small circle. In particular, when the Christian gets to T, so does the Lion.

To justify this Claim, all we have to show is that the length of the SC-arc of the big circle is precisely the length of the OB-arc of the small circle (with centre O'). This is immediate from the two facts that a) the line OS is a tangent of the small circle, so the angle $BO'O$ is twice as large as the angle COS, and b) the radius of the big circle is twice the radius of the small one.

Conclusion. *The Lion wins the LC game.* □

This solution shows that in the first argument we were too hasty to assume that the best strategy for the Lion is to run right towards the Christian. In fact, as we have seen, the Lion can do better by adopting a less obvious strategy: he tries to cut off the path of the Christian by running towards his future position.

Some readers may well suspect that we are trying to fool them, and the Christian can escape by simply reversing his direction along the boundary of the circle. This is not the case at all: as the Lion is on the radius leading to the Christian, for the Lion it makes no difference in which direction the Christian runs. By reversing his direction along the boundary the Christian gains nothing, since according to his strategy the Lion just stays on the radius, and so runs towards the same side as the Christian. No matter how often the Christian changes his direction while running at full speed along the boundary, he will be caught in exactly the same time as before. And slowing down just leads to a swifter end.

This is where the problem and its solution rested for about twenty years: the 'stay-on-the-radius' strategy is a quick win for the Lion. This is nice, but rather boring: for a mathematician it is not worth a second glance. Then, in the 1950s, a thunderbolt struck, when Besicovitch found the following beautiful argument. It is not clear what prompted Besicovitch to consider the problem again: it is quite possible that he wanted to mention it in an after-dinner talk to mathematics undergraduates in his College, Trinity.

Third Solution: Run Along a Polygonal Path of Infinite Length.
In this solution we describe a strategy for the Christian. Trivially, we may assume that in the starting position the Christian is in $C_1 \neq O$, the Beast is in $B_1 \neq C_1$ and the length $\overline{OC_1}$ is r_1, where $0 < r_1 < 1$.

Claim. Suppose that there are positive numbers $t_1, t_2, \ldots$ such that $\sum_i t_i$ is infinite but $\sum_i t_i^2 < 1 - r_1^2$. Then the Christian can escape.

To show this, split the (infinite) time into a sequence of intervals, of lengths $t_1, t_2, \ldots$. We shall 'review the situation' at times $s_i = \sum_{j=1}^{i-1} t_j$, $i = 1, 2, \ldots$, calling the time period between s_i and $s_{i+1} = s_i + t_i$ the *i*-th *step*. For convenience, set $t_0 = r_1$ so that $\sum_{i=0}^{\infty} t_i^2 < 1$.

Suppose that at time s_i the Christian is at point $C_i \neq O$, the Beast in $B_i \neq C_i$, and C_i is at a distance $r_i = \overline{OC_i}$ from the centre, where $r_i^2 = \sum_{j=0}^{i-1} t_j^2 < 1$. (Since the Christian starts at C_1, this is consistent with our earlier assumption. The condition $C_i \neq O$ is utterly unimportant: its only role is to make the description below slightly easier.) Let ℓ_i be the line through O and C_i. At the i-th step the Christian runs for time t_i in a straight line perpendicular to ℓ_i, in the direction that takes him away from B_i as much as possible. To be precise, if B_i is not on ℓ_i, but in the interior of one of the half-planes bounded by ℓ_i then B_i runs away from this half-plane, otherwise (i.e. if B_i is on ℓ_i) either direction will do. During this step the Christian runs away from the line ℓ_i; as the Lion starts either on ℓ_i or on its 'wrong' side, he has no chance of catching the Christian in this time step (see Figure 4). In particular, $C_{i+1} \neq B_{i+1}$ and $C_{i+1} \neq O$.

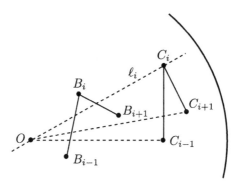

Fig. 4. The polygonal path of the Christian.

How far is the Christian from the centre at time s_{i+1}? By Pythagoras' theorem, the square of this distance $\overline{OC}_{i+1}$ is precisely $r_i^2 + t_i^2 = \sum_{j=0}^{i} t_j^2 = r_{i+1}^2 < 1$. Hence the polygonal path $C_1 C_2 \ldots$ of *infinite* length the Christian runs along remains in the disc D, and the Christian is not caught during this run, completing the proof of our Claim.

Finally, the Claim gives a winning strategy for the Christian, since it is easy to choose a sequence $t_1, t_2, \ldots$ satisfying the conditions in our Claim: e.g. we may take $t_i = 1/(i+r)$ for r large enough, since $\sum_{i=1}^{\infty} 1/i = \infty$ and $\sum_{i=1}^{\infty} 1/i^2$ is finite.

Conclusion. *The Christian wins the LC game.* □

Clearly, this is 'where the buck stops'. Bessie's solution is indeed correct: using his polygonal path strategy, the Christian can indeed escape, no matter what the Lion does. The first 'solution' collapsed since we had no right to assume that the Lion rushes straight at the Christian; the second 'solution' collapsed since it was still based on the unjustified assumption that the best strategy for the Christian is to run along the boundary. The third, *correct* solution shows that, like a boxer on the ropes, in the second solution the Christian puts himself at a disadvantage by restricting his movements.

There is an obvious variant of the strategy in the third solution, which is slightly 'better' for the Christian: in the i-th step, he can run in a direction perpendicular to the $B_i C_i$ line, starting in the direction which initially takes him closer to O. Unless the points B_i, C_i and O are collinear, this is better for the Christian in the sense that he stays further away from the boundary.

3 Variants

There are numerous variants of the LC game: here we shall mention only a few, leaving the exact formulations to the reader. Let us start with a question which must have occurred to the reader some time ago.

1. Does the Shape of the Arena (Within Reason) Matter? *Could the Lion win in an oval arena, the kind the Romans really had? Or in a triangular arena? Could the Lion drive the Christian into a corner of the triangle and then devour him?*

No doubt, a reader who has paid a little attention to the strategies we have given will see through this question in an instant.

Turning to less trivial variants, here are two results proved by Croft [4].

2. Birds Catching a Fly. *There are some Birds and a Fly in the d-dimensional closed unit ball, each with the same maximum speed. What is the minimal number of Birds that can catch the Fly?*

Fig. 5. Birds catching a Fly.

What Bessie's solution of the LC game tells us is that for $d = 2$ one bird is not enough; it is easy to see that two birds suffice. In general, $d - 1$ birds do not suffice, but d do.

3. Uniformly Bounded Curvature. *If the Christian is forced to run along a curve of uniformly bounded curvature, then the Lion can win the LC game.*

Roughly, if the Christian cannot change direction arbitrarily fast, then the Lion can catch him.

What happens if we have many Lions and one Christian, but the game is played on the entire plane rather than in a bounded arena? The answer was given by Rado and Rado [9] and Janković [5].

4. Many Lions in the Plane. *Finitely many Lions can catch a Christian in the plane in finite time if and only if the Christian is in the interior of the convex hull of the Lions.*

Finally, here is a problem which is still open.

5. Two Lions on a Golf Course. *Can two Lions catch the Christian on a golf course with finitely many rectifiable lakes?*

Needless to say, the assumptions are that neither the Christian nor the Lions are allowed to step into the lakes, and the boundaries of the lakes are 'nice' in a technical sense (see Figure 6).

Fig. 6. Lions trying to catch a Christian.

4 Mathematical Formalism

Having read three 'solutions' of the LC game, the reader is entitled to wonder whether we actually know what a 'winning strategy' really means. If the game is played using alternate moves, then there is no problem: in every time step, knowing the position of the game, the player has to decide what to do next. In particular, a winning strategy amounts to a choice of moves ending in a win, no matter what the opponent does. However, if the game is played in continuous time, we have to be more careful. Indeed, can we say precisely what a strategy is in the continuous LC game? To answer this question precisely, we need some definitions. We shall write $|x|$ for the modulus of a number x and also for the length of a vector x. In particular, if $x, y \in D$ then $|x - y|$ denotes the distance between the points x and y.

Suppose the Lion starts at x_0 and the Christian at y_0, and both have maximum speed 1. A *Lion path* is a map f from $[0, \infty)$ to the unit disc D such that $f(0) = x_0$ and $|f(t) - f(t')| \leq |t - t'|$ for all times $t, t' \geq 0$. (Such a map f is said to be 'Lipschitz', with constant 1.) Similarly for a Christian path. When the Lion follows a path f then at time t he is at $f(t)$, and when the Christian follows a path g then at time t he is at $g(t)$.

Let $\mathcal{B}$ be the set of Lion paths (the paths of the beast) and $\mathcal{C}$ the set of the Christian paths. Then a *strategy* for the Christian is a map Φ from $\mathcal{B}$ into $\mathcal{C}$ such that if $f_1, f_2 \in \mathcal{B}$ agree up to time t_0 (i.e. $f_1(t) = f_2(t)$ for all $0 \leq t \leq t_0$) then $\Phi(f_1)$ and $\Phi(f_2)$ agree on $[0, t_0]$. This 'no lookahead' condition tells us that $\Phi(f)(t)$ depends only on the restriction of f to the interval $[0, t]$. A Lion strategy $\Psi : \mathcal{C} \to \mathcal{B}$ is defined similarly. A Christian strategy Φ is *winning* if $\Phi(f)(t) \neq f(t)$ for every path $f \in \mathcal{B}$ and for every $t \geq 0$. And a Lion strategy Ψ is *winning* if for every path $g \in \mathcal{C}$ of the Christian there is a time $t \geq 0$ such that $\Psi(g)(t) = g(t)$.

Are these the 'correct' definitions? A moment's thought tells us that they are, since we want to allow strategies *without delay*, like the 'curve of pursuit' and 'stay on the radius' strategies of the Lion in the first two 'solutions' above.

Note that these definitions make sense in more general circumstances. For instance, the arena need not be the disc: any set in the plane or 3-dimensional space would do. In fact, so would any metric space (a space with a sensible 'distance' function). Consequently, pursuit–evasion games like the LC game make good sense in these general metric spaces. As we wish to give the Lion a chance to catch the Christian, we shall always assume that the playing field (our metric space) contains a path from the Lion to the Christian.

Having defined what we mean by a winning an LC-type pursuit–evasion game, let us ask a question our readers may find surprising. We know that in the disc Bessie's strategy is a win for the Christian;

but could it happen that the Lion also has a winning strategy?

Surely many readers would agree that this question is not only surprising but also downright crazy. How on earth could both have a winning strategy? Of course not, we say loud and clear:

if both had winning strategies then with each playing his own winning strategy we would get a game which is a win for the Christian and also a win for the Lion — a blatant contradiction.

A little thought should tell us that, once again, we have been too hasty: this 'argument' is sheer nonsense. Indeed, how can *both* play their winning strategies? Suppose that the outcome of a game in which both play their winning strategies, Φ and Ψ, is a Lion path f and a Christian path g. Then $\Phi(f) = g$ and $\Psi(g) = f$; in particular, $\Psi(\Phi(f)) = f$, i.e. f is a fixed point of the composite map $\Psi \circ \Phi$ mapping $\mathcal{B}$ into itself. *But why should the composite map $\Psi \circ \Phi$ have a fixed point at all?* There is no reason why it should.

Thus, in a general pursuit–evasion game, there are two basic questions. Does the Lion have a winning strategy? Does the Christian have a winning strategy? Can all four conceivable combinations of answers occur?

If the moves in our game alternate, say, if the moves take place at times 1, 2, etc., with the Christian moving at odd times and the Lion at even times, then there is no problem with playing a strategy against another. Thus in this case it cannot happen that both players have winning strategies. What gives the present continuous problem an entirely different complexion is that whoever *plays his strategy* can react *instantaneously* to the move (really 'path' or 'trajectory') of the other. This is certainly the right definition if we want to allow strategies like the Lion's 'stay on the radius' strategy, in which the Lion is shadowing the move of the Christian, instantly reacting to any change of speed or direction of his prey.

There are also several other natural questions. For example, are there 'nice' winning strategies? The most obvious way a strategy can be 'nice' is that it is *continuous*: it maps 'nearby' paths into 'nearby' paths. (Formally, a Christian strategy $\Phi : \mathcal{B} \to \mathcal{C}$ is continuous if for every $f_0 \in \mathcal{B}$ and $\varepsilon > 0$ there is a $\delta > 0$ such that if $f_1 \in \mathcal{B}$ is such that $|f_0(t) - f_1(t)| < \delta$ for every t then $|\Phi(f_0)(t) - \Phi(f_1)(t)| < \varepsilon$ for every t. And similarly for a Lion strategy.)

Also, what happens if we play the *bounded-time game*, i.e. the entire game must run its course by a fixed time T? In this version the Christian wins if he can stay alive up to time T. And what happens if we postulate that the playing field is 'nice'?

If everything goes as we 'feel' it should, then our mathematical formalism is rather wasted. However, the results in the final section show that this is far from the case: there are several unexpected twists.

5 Results and Open Problems

In this final section we shall give a selection of results from a recent paper by Bollobás, Leader and Walters [3], and state some open problems.

We shall play the bounded-time game on a playing field which is just about as nice as possible: a *compact* metric space, so that every sequence has a subsequence converging to a point of the space, like a closed interval, a closed disc or a closed ball.

What about the following two statements?

1. At least one of the players has a winning strategy.
2. At most one of the players has a winning strategy.

Certainly, both statements feel true. In fact, by considering the discretised version of the game, one can show that the first statement is true in the best of all possible worlds: in a compact metric space, at least one of the players has a winning strategy for the bounded-time game. It turns out that not even this assertion is trivial to prove; for a proof, see [3].

And what about the second statement, which feels just as much true? Surprisingly, this statement is false.

A Game in Which Both Players Have Winning Strategies. *Let the playing field be the closed solid cylinder*

$$D \times I = \{(a, z) : a \in D \text{ and } 0 \leq z \leq 1\},$$

with the distance of two points $(a, z), (b, u) \in D \times I$ *defined to be*

$$\max\{|a - b|, |z - u|\}.$$

At the start, C is at the centre of the top of the cylinder (a unit disc), and B is at the centre of the bottom. Then both players have winning strategies.

Proof. Having defined this game, it is very easy to justify these assertions. Indeed, the Christian can win if he can make a small move which takes him away from being exactly above the Lion, since from then on he can simply run the Bessie strategy, ignoring the height, and so survives for ever. But can he get away from above the Lion in time $t_0 < 1/2$, say? Unexpectedly, this can be done very easily. We encourage the reader to find a simple argument for this.

And how can the Lion win? That is even easier. He keeps his disc coordinate the same as the Christian, and increases his height with speed 1 until he catches him. This will happen by time 1. Note that the Lion makes use of the fact that the distance on the solid cylinder is not the usual Euclidean distance, but the so-called ℓ_∞ *distance*, the maximum of the distances in the disc D and the interval I. □

Note that the simple strategies in the proof above are examples of strategies which cannot be played against each other.

Let us return to the original Christian and Lion game in the closed unit disc D. Bessie's strategy is a winning strategy for the Christian, and it turns out that it can be discretised to show that the Lion does *not* have a winning strategy. But is Bessie's strategy continuous? By considering the positions in which O, B and C are collinear, we can see that it is not. In fact, considerably more is true.

Continuity in the Original Lion and Christian Game. *In the original game, neither player has a continuous winning strategy.*

Furthermore, with the Lion starting in the origin, for any continuous strategy of the Christian, there is a Lion path catching the Christian by time 1.

Proof. As we know that the Lion does not have a winning strategy, we have to prove only the second assertion. For this, we need a classical result from topology, *Brouwer's fixed point theorem,* stating that every continuous map $\varphi : D \to D$ has a fixed point, i. e. a point $x \in D$ such that $\varphi(x) = x$ (see, e.g., [1], p. 216).

Let then $\Phi : \mathcal{B} \to \mathcal{C}$ be a continuous Christian strategy. For every $z \in D$, let h_z be the constant speed straight path from 0 to z, reaching z at time 1, i.e. set $h_z(t) = tz$ (assuming, as we do, that the origin is the centre of D.) Then $z \mapsto h_z$ is a continuous map from D into $\mathcal{B}$. Consequently, $z \mapsto \Phi(h_z)(1)$ is a continuous function mapping D into D. By Brouwer's fixed point theorem, this map has a fixed point $z_0 \in D$: $\Phi(h_{z_0})(1) = z_0$. Hence, if the Lion follows h_{z_0}, and the Christian plays using his strategy Φ then the Lion catches the Christian at time 1. □

Rather curious phenomena can occur if the arena is not compact, even if otherwise it is as nice as possible. For example, Alexander Scott noted that in the LC game played on the *open* interval $(0, 1)$ *both* the Lion and the Christian have winning strategies. Indeed, suppose the Lion starts at 2/3 and the Christian at 1/3. Then

$$f(t) \mapsto \Phi(f)(t) = f(t)/2$$

is a winning strategy for the Christian, and

$$g(t) \mapsto \Psi(g)(t) = \max\{2/3 - t, g(t)\}$$

is a winning strategy for the Lion.

However, as we mentioned earlier, it was shown in [3] that a bounded-time LC game played on a compact field cannot be too pathological: at least one of the players has a winning strategy.

In a bounded-time LC game played on a compact field at least one of the players has a winning strategy.

Finally, let us leave the reader with two open questions which arise naturally from this result.

1. Is there a bounded-time LC game in which neither player has a winning strategy?

2. Is there an unbounded-time LC game played on a compact field in which neither player has a winning strategy?

For other results and questions, the reader is referred to the original paper [3].

Acknowledgement. I am grateful to Gabriella Bollobás for her pen drawings.

References

[1] Béla Bollobás, *Linear Analysis: An Introductory Course*, second edition. Cambridge University Press, Cambridge (1999), xii+240 pp.
[2] Béla Bollobás, *The Art of Mathematics — Coffee Time in Memphis*. Cambridge University Press, New York (2006), xvi+359 pp.
[3] Béla Bollobás, Imre Leader, and Mark Walters, *Lion and man — can both win?* Preprint. http://arxiv.org/abs/0909.2524, 24 pages (September 14, 2009)
[4] Hallard T. Croft, "Lion and man": a postscript. *Journal of the London Mathematical Society* **39**, 385–390 (1964)
[5] Vladimir Janković, About a man and lions. *Matematički Vesnik* **2**, 359–361 (1978)
[6] John E. Littlewood, *Littlewood's Miscellany*. Edited and with a foreword by B. Bollobás. Cambridge University Press, Cambridge (1986), vi+200 pp.
[7] Paul J. Nahin, *Chases and Escapes — The Mathematics of Pursuit and Evasion*. Princeton University Press, Princeton (2007), xvi+253 pp.
[8] C.C. Puckette, The curve of pursuit. *Mathematical Gazette* **37**, 256–260 (1953)
[9] P.A. Rado and R. Rado, *Mathematical Spectrum* **7**, 89–93 (1974/75)

Three Mathematics Competitions

Günter M. Ziegler

Abstract. The development of mathematics is based on cooperation, collaboration, joint efforts, and joint work. Nevertheless, mathematicians compete. Indeed, they compete in many different ways, in very different races, and in diverse arenas. In this little contribution, I want to tell you about three different types of mathematics competitions that are different from the IMO.

1 Computing π

Much older than International Mathematical Olympiads is the "compute as many different digits of π as you can" competition, which was started in antiquity. The first person to compute more than 10 decimal digits correctly was (as far as we know) Al-Kashi, in 1429; Machin computed 100 digits in 1706, and in 1949 Smith and Wrench obtained more than 1000 digits — using a desktop calculator.

Why would they do this? Of course, any university press office is happy if they can announce a world record. For the mathematicians, however, it is a kind of sport, a competition. But a more serious reason is that massive, record-breaking computations demonstrate the extent to which theory has made progress, and which of the many wonderful formulas for π [20] can really be *used* for a computation. Whoever wants to compute, say, a million digits of π can't be content with 18-th century formulas. There is another reason for the record races: they are good for testing computers, hardware as well as software. Accordingly, all such computations are supposed to be

---

Günter M. Ziegler
Fachbereich Mathematik und Informatik, Freie Universität Berlin, Arnimallee 2, 14195 Berlin, Germany. e-mail: ziegler@math.fu-berlin.de

done in two different ways, and the results are checked against each other, and against the result of the previous record computation.

For a long time, the world record was held by Yasumasa Kanada and his team, which, in November 2002, after more than 600 hours of computation, obtained the first 1.2 trillion ($1.2 \cdot 10^{12}$) digits of π on a Hitachi supercomputer SR8000/MPP with 144 processors. This record held until August 17, 2009: Daisuke Takahashi used the supercomputers at Tsukuba University with a peak performance of 95 trillion floating point operations per second (95 teraflops) for a computation which still took 73 hours and 36 minutes — to compute 2.577 trillion digits of π.

This record, however, held only for 136 days: on December 31, 2009, the French programmer Fabrice Bellard announced the computation of the first 2 699 999 990 000 decimal digits of π — that's nearly 2.7 trillion digits, and over 123 billion more than the previous record. The Japanese may have been surprised by this announcement, and they may not really have liked it, because Bellard did *not* use a super-expensive supercomputer. Instead, he used a single PC that cost less than 2000 Euro. He did, of course, use an advanced formula for the computation:[1]

$$\frac{1}{\pi} = \frac{3}{\sqrt{40\,020}} \sum_{n=0}^{\infty} (-1)^n \binom{6n}{3n,n,n,n} \frac{545\,140\,134\,n + 13\,591\,409}{640\,320^{3n+1}}.$$

This remarkable formula, which yields 14 new digits of π with every single summand, was discovered by the legendary Chudnovsky brothers in 1984. They live together in New York (David assisting Gregory, who suffers from the autoimmune muscular disease *myasthenia gravis*), and they both hold professorships at the NYU Polytechnic Institute. The Chudnovsky brothers are famous not only for their formulas which can be used to compute π, but also for their "home made" supercomputer which they used to compute the first one billion digits of π. This computation held the world record from 1989 to 1997. See [16] for a remarkable rendition of the Chudnovsky's story. (The Chudnovsky brothers may also have inspired the 1998 movie "Pi", directed by Darren Aronofsky ...)

The Chudnovskys' amazing formula did not come out of the blue. It was inspired by earlier formulas due to the Indian genius Srinivasa Ramanujan, who, for example, came up with this formula:

$$\frac{1}{\pi} = \frac{2\sqrt{2}}{9\,801} \sum_{n=0}^{\infty} \binom{4n}{n,n,n,n} \frac{26\,390\,n + 1\,103}{396^{4n}}.$$

---

[1] The "multinomial" coefficient $\binom{n}{k_1,k_2,k_3,k_4} := \frac{n!}{k_1!k_2!k_3!k_4!}$ denotes the number of partitions of a set of n elements into four subsets of sizes k_1, k_2, k_3, and k_4, where $n = k_1 + k_2 + k_3 + k_4$.

Such formulas (see e.g. [20]) are deeply rooted in the theory of modular forms — a theory that Ramanujan did not have available in his time. The life of Ramanujan has also been the subject of literature, including the biography "The Man Who Knew Infinity", by Robert Kanigel (1991), the novel "The Indian Clerk", by David Leavitt (2007), and the theater play "A Disappearing Number", by Simon McBurney and the Théâtre de Complicité company (2007).

For his recent record, Bellard used the Chudnovskys' formula [5], and he certainly also coded it well — so that it could run on a standard desktop PC (with a Core i7 CPU at 2.93 GHz). Technical details are given in [1]. His calculation, including verification, took 131 days, so he must have started the computation just a few days after the Japanese 2009 world record had been announced.

But of course the race continues. Bellard's record stood for seven months and three days: on August 2, 2010, Alexander J. Yee and Shigeru Kondo announced the computation of 5 trillion digits of π, the next new world record, again on a single desktop computer [22]. And by the time you read this, this record may not stand any more ...

2 Mathematician vs. Mathematician

This is the story of a remarkable public competition between two mathematicians that took place in the year 1894. In mathematics, there are sometimes races for the solution of a mathematical problem. Such a fight will become public in the best (or worst) case only after the race, as in the case of the fierce, unfair, and destructive fight between Newton and Leibniz about priority in the invention of calculus. The 1894 competition that I want to tell you about was not primarily about mathematics, but about chess, the most mathematical of all strategic games. Chess is pure logic. It is logical thinking that counts, as well as strategy and the proper evaluation of positions; thus, chess competitions are an arena for mathematicians.

May I introduce the competitors to you? The first was *Wilhelm Steinitz*, born in Prague in 1836. He came to Vienna in 1858 in order to study mathematics. At that time he earned his living as a parliamentary newspaper reporter for the "Österreichische Constitutionelle Zeitung", but he found out rather soon that it was much easier to earn money by playing chess at Viennese coffee houses. Steinitz played a lot of chess (and neglected his mathematics studies, as we are forced to assume). In 1862 he played his first international tournament, in London. I don't know whether he ever graduated from his mathematics studies — but his approach to chess gives the mathematician away. Steinitz is today seen as a revolutionary of chess theory: we owe to him the "scientific approach" to chess, the systematic search for rules and patterns. And this led him to success. He practiced "theoria cum praxi" (to quote a motto that the mathematician Gottfried Wilhelm Leibniz proposed in March of 1700 on the occasion of the founding of the Prussian Academy of Sciences), and won one tournament after another. In a fierce battle in London in 1866 (which ended 8:6 — there was not a single draw in the 14 games played), he defeated Adolf Anderssen from Prussia, who had also studied mathematics. Anderssen was an exponent of the "romantic", attack-at-all-costs style of chess, and until then had been regarded as the unofficial world chess champion. From this point on Steinitz was seen as the best chess player in the world. From 1866 up to the world championship fight in 1894 — that is, for twenty-eight years — he dominated the world of chess. In 1886, aged fifty, he defeated Johannes Hermann Zukertort from Poland. From this point on he was the first official chess world champion.

The second competitor was *Emanuel Lasker*, a German jew, who was born in 1868 in Berlinchen (Neumark, which is today in Poland). He was a brother-in-law of the German poet Else Lasker-Schüler. Lasker started to study mathematics in Berlin in 1889, but one year later he moved to Göttingen. In the same year his chess career started with a victory at the Main Tournament in Breslau (Wrocław, Poland). At some point after this, chess must have dominated mathematics in his life: Emanuel Lasker interrupted his mathematics studies in 1891, moved first to London and then in 1893 to the US.

One year later, in 1894, the decisive duel "Mathematician vs. Mathematician" took place: the 25-year-old Lasker against the 58-year-old Steinitz. You are welcome to voice your sympathies — for the distinguished senior master or for the youthful challenger. Supporters for both sides collected prize money amounting to 3000 US Dollars, of which the winner was to get 2250, and the loser the rest. The *New York Times* reported that there was lots of "excitement in chess circles all around the world". The contest was to take place first in New York, then in Philadelphia, and then in Montreal, until one of the contestants had won 10 matches.

The fight starts on March 15, 1894. Lasker wins the first game, Steinitz the second, Lasker the third, and Steinitz the fourth. Then there are two draws. The score is now 2:2, since draws don't count. The fight is dramatic, and the odds change several times. Eventually Lasker wins games number 15 and 16, and thus takes a 9:4 lead. He is missing one more victory, but then Steinitz strikes back, winning the 17-th game in the "great style of his heyday". The game is said to be the best in the whole competition. Can Steinitz still turn the tables? Will the old man succeed in the end? Lasker still can't win the next match, despite having an apparent advantage throughout the game.

The two competitors were similar in their styles: they played the modern, systematic positional chess style introduced by Steinitz. But in addition Lasker may have used psychology to his advantage — he was not interested in the scientifically correct move, but only in the one that was most annoying to his opponent. At least this was claimed by one of his badly inferior competitors.

Did the public hear about the competition? Yes, it seems so: the *New York Times* reported on all the games in detail. Did they view this as a fight "Mathematician vs. Mathematician"? That I don't know.

But then, finally, on May 26, 1894, Lasker is pertinacious in winning the 19-th and last game, and the final score looks quite unambiguous: Lasker wins the championship 10:5 (with four games ending in a draw).

> Lasker, therefore, is champion of the world, and deserves to be congratulated upon his success, inasmuch as he has beaten his man fairly and decisively, and thereby justified the confidence which was placed in him by his backers,

the *New York Times* reports the next day. Lasker, whom the *New York Times* termed "the Teuton" after this victory, is also the first and so far only German chess world champion.

Two and a half years later, at the turn of the years 1896/1897, a rematch takes place in Moscow. This time Lasker wins much more clearly, 10:2 (with five draws), and keeps the title of the chess world champion for 27 years, until 1921, longer than anyone else (up to now).

As a mathematician, I claim Lasker as "one of us": he was not a chess world champion with unfinished mathematics graduate studies, but someone who *was* a mathematician and wanted to be exactly that. Indeed, after his second fight against Steinitz he retired from chess for a while and continued his mathematics studies, first in Heidelberg and then in Berlin. In 1900 he completed his Ph.D. at the University of Erlangen as a student of Max Noether, the father of Emmy Noether. His dissertation "Über Reihen auf der Convergenzgrenze" was only 26 pages long. It was published in 1901. Four years later, in 1905, a long and important paper of his appeared in the journal *Mathematische Annalen*. The paper was on commutative algebra and introduced the concept of "primary decomposition". This line of investigation was later continued by Emmy Noether. Apparently, Lasker hoped for an academic career in mathematics, but since he could not find a suitable position in Germany, England, or the US, he had to continue playing chess. Perhaps Lasker was a real-life predecessor to the ingenious piano virtuoso Frantisek Hrdla in Wolfgang Hildesheimer's story "Guest Performance of an Insurance Agent" (1952), who really wanted to be an insurance agent, but whose dominating father kept him from taking up the profession of his dreams ...

What was Lasker's dream profession? Albert Einstein writes in the preface of a biography [11] about Lasker, whom he had met in Berlin and come to know well on their common walks:

To my mind, there was a tragic note in his personality, despite his fundamentally affirmative attitude towards life. The enormous psychological tension, without which nobody can be a chess master, was so deeply interwoven with chess that he could never entirely rid himself of the spirit of the game, even when he was occupied with philosophic and human problems. At the same time, it seemed to me that chess was more a profession for him than the real goal of his life. His real yearning seems to be directed towards scientific understanding and the beauty inherent only in logical creation, a beauty so enchanting that nobody who has once caught a glimpse of it can ever escape it.

Spinoza's material existence and independence were based on the grinding of lenses; chess had an analogous role in Lasker's life.

It is interesting to note that Lasker stepped onto Einstein's territory, so to speak, by publishing a paper critical of Einstein's special theory of relativity, where he questioned the hypothesis that the speed of light is constant in a vacuum.

A strange man, I thought (...) truly a double talent of unusual degree.

(This quote, however, is not Einstein about Lasker, but Hildesheimer about Hrdla.)

3 Packing Tetrahedra

How densely can one pack equal-sized regular tetrahedra in space? This question was posed by Hilbert as part of the 18-th of his famous problems presented at the 1900 International Congress of Mathematicians in Paris [13]:

I point out the following question, related to the preceding one, and important in number theory and perhaps sometimes useful in physics and chemistry: how can one arrange most densely in space an infinite number of equal solids of given form, e.g. spheres with given radii or regular tetrahedra with given edges (or in prescribed position), that is how can one fit them together so that the ratio of the filled to the unfilled space may be as great as possible?

But indeed the story starts much earlier. The greek philosopher Aristotle claimed that there was a perfect packing in which the tetrahedra fill space completely, without gaps — a 100% packing. This is not true, but the full truth is even worse. Aristotle writes:[2]

---

[2] *De Caelo* III, 306b; quoted from Majorie Senechal's prize winning paper "Which Tetrahedra Fill Space?" [17].

It is agreed that there exist only three plane figures that can fill a place, the triangle, the quadrilateral, and the hexagon, and only two solid bodies, the pyramid and the cube.

Clearly the "figures" refer to *regular* polygons or polyhedra only, and the "pyramid" in question is a regular tetrahedron. As for filling space with regular tetrahedra, Aristotle not only claims that this is possible, he says of it that "it is agreed". Well known? Maybe, but it is not true! However, if the great Aristotle refers to this as known and agreed, then it will take time before anyone seriously dares to question it ....

The error stood for nearly 1800 years, until the German Johannes Müller (1436–1476), known as Regiomontanus, a father of modern trigonometry, uncovered it. His manuscript "De quinque corporibus aequilateris quae vulgo regularis nuncupantur: quae videlicet eorum locum impleant corporalem & quae non. contra commentatorem Aristotelis Averroem"[3] seems to be lost, so we do not know what he wrote in detail — but the title gives a clear indication. That Aristotle's claim is false can be verified using a carefully constructed cardboard model, or, more easily (and more reliably), using a bit of trigonometry and a pocket calculator: the dihedral angle at each edge of a regular tetrahedron is $\arccos \frac{1}{3} \approx 70.529°$, and thus just a bit less than a fifth of $360°$. But of course this trigonometry, and the pocket calculator, were not available to Aristotle and his contemporaries. Apparently Regiomontanus could do the math: he had the trigonometry at hand.

But if 100% of the space cannot be filled, how densely can one pack tetrahedra? That one can pack cubes perfectly can be seen, for example, in any package of sugar cubes. Equal-sized spheres can fill space up to a fraction of $\frac{\pi}{\sqrt{18}} \approx 74.05\%$ — this was the Kepler Conjecture, made in 1611, which was settled in 1998 by Thomas C. Hales together with his student Samuel Fergusson [10], based on extensive computer calculations — see [12] and [18].

But tetrahedra? How dense can a "sand" whose grains are equal-sized, regular tetrahedra be? This problem is rather easy to solve if you assume that all the tetrahedra have the same orientation in space, and moreover that their centers form a lattice. Then the densest packing fills only $\frac{18}{49} \approx 36.73\%$ of three-dimensional space: see Figure 1.

If you do not require the lattice structure, then things get much more complicated (the same was also true for the case of spheres, for which Gauß had solved the problem of a lattice packing). But if you also allow the tetrahedra to be individually rotated in space, then it gets really complicated. This is *Tetrahedron Tetris*©?: you can and should try to rotate the tetrahedra cleverly so that they fit into the gaps left by the others. But how dense a packing can one achieve?

---

[3] "On the five like-sided bodies, that are usually called regular, and which of them fill their natural place, and which do not, in contradiction to the commentator of Aristotle, Averroës".

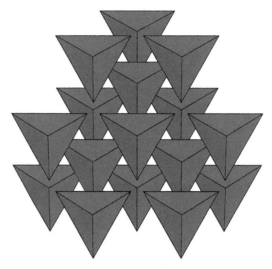

Fig. 1. The densest tetrahedron lattice packing. Graphics from [6].

It is only recently that this problem has moved into the focus of research — and become the object of a competition that involved scientists from quite different disciplines. My rendition of the story owes a lot to a *New York Times* report by Paul Chang [2], which tells a lot of the story (but by far not all of it, as I am told by scientists close to the race).

The starting signal for the current competition was given in 2006 by John H. Conway, a legendary Princeton geometer and group theorist, together with Salvatore Torquato, a colleague of his from the chemistry department. Together they obtained a remarkably bad result, which they published in the *Proceedings of the National Academy of Sciences*: they could not fill more than 72% of space with equal-sized regular tetrahedra — this is worse than the optimal sphere packing!

This seemed incredible to Paul M. Chaikin, an NYU physicist: he bought large packs of tetrahedral dice (as used for the game "Dungeons & Dragons") and let high school students experiment with them. With a bit of wobbling and shaking of tetrahedra in large containers, they got a percentage significantly greater than 72%. But of course such physical experiments would not be accepted as proofs in a mathematics community — since, for example, the plastic tetrahedra used here have slightly rounded edges and vertices, and hence are not ideal tetrahedra. Does that make much of a difference? That is hard to tell!

At the same time in Ann Arbor, Michigan ... the mathematician Jeff Lagarias challenged his Ph.D. student Elizabeth Chen: "You've got to beat them. If you can beat them, it'll be very good for you." Chen got going, analyzed lots of possible local configurations, and in August 2008 presented a packing with a remarkable 78% density [3]. At first, Lagarias wouldn't even believe her!

A bit later ... at the same university, but at the chemical engineering department, Sharon C. Glotzer became interested in tetrahedron packings: she and her colleagues wanted to find out whether, upon shaking, tetrahedra would fall into the crystalline structures that they knew from liquid crystals. In order to find this out, they wrote a computer program to simulate the shaking and rearrangement of tetrahedra — and found a complicated, "quasi-crystalline" structure that consists of lattice-like repetitions of a basic configuration of 82 tetrahedra. Complicated, but dense: 85.03%! While these results were on the way to publication in *Nature* [9], competitors emerged: Yoav Kallus, Simon Gravel, and Veit Elser from the Laboratory of Atomic and Solid-State Physics at Cornell University found a much simpler packing that is built up from repetitions of a simple configuration of 4 tetrahedra [14]. (It is not at all clear why this simple configuration did not turn up in the simulations by Glotzer et al.) Density: 85.47%.

But the race went on ... Shortly before Christmas 2009, Salvatore Torquato and his Ph.D. student Yang Jiao achieved a density of 85.55%: they analyzed the Cornell solution and managed to improve it slightly [19]. Was this the end of the race?

No! On December 26, 2009, Elizabeth Chen struck back: her preprint, submitted to the arXiv just after the end of the year (and written jointly with Sharon Glotzer and Michael Engel from the chemical engineering department) describes a further improvement of the Cornell crystal; it was obtained by a systematic optimization ansatz [4]. Density: $\frac{4000}{4671} \approx 85.6348\%$. And this, nearly one year later (November 2010), still seems to be the current record.

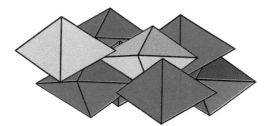

Fig. 2. An optimized configuration of $N = 16$ tetrahedra that repeats a configuration of two double-tetrahedra. Graphics from [4].

Where is the finishing line for the race? I don't know, of course. And as far as I know there are currently no good estimates at all for the distance to the optimum. Perhaps 85.6348% is optimal, perhaps there are much better packings. Now we have to look for upper bounds, and these cannot be obtained via constructions but require quite different tools. Perhaps estimates such as those used for the Kepler problem (see Lagarias [15] and Henk & Ziegler [12]) can help, and perhaps they can't.

However, I would now expect a race that starts at the other end, at the 100% mark: Who can show that a packing of equal-sized regular tetrahedra cannot fill more than 95% of three-space? All that is proved at the moment (November 2010) seems to be that the density cannot be more than 99.999 999 999 999 999 999 974%, according to Gravel et al. [8].

How Does It Feel?

How does it feel to compete in mathematics? IMO participants know. (Others may get a glimpse from George Csicsery's splendid documentary [7] about the US team at the 2006 IMO. A film by Oliver Wolf about the 2009 IMO was finished in the summer of 2010 [21].) As far as I can tell, chess competitions are more physically demanding, but research competitions are much more collaborative, as you can see from the tetrahedron race. There are many more races going on. Some of them are treated with a certain amount of secrecy — see for example the sailing competition "America's Cup", which has over the years turned into a mathematicians' competition, at least to some extent. Competitors include, for example, the mathematician Alfio Quarteroni from the École Polytechnique Fédérale de Lausanne, who, along with his team, is deeply involved in the design and optimization of the Swiss yacht Alinghi, which won the competition twice. But most of the races are much less a competition than a collaborative effort, and anyone who is looking for an intellectual challenge (and perhaps for a chance to contribute, and to prove his or her talent and capability), is welcome to join in. Do it!

References

[1] Fabrice Bellard, *Computation of 2700 billion decimal digits of Pi using a desktop computer.* Preprint (4th revision). http://bellard.org/pi/pi2700e9/pipcrecord.pdf, 11 pages (February 11, 2010)

[2] Kenneth Chang, Packing tetrahedrons, and closing in on a perfect fit. *New York Times.* http://www.nytimes.com/2010/01/05/science/05tetr.html (January 4, 2010)

[3] Elizabeth R. Chen, A dense packing of regular tetrahedra. *Discrete & Computational Geometry* **40**, 214–240 (2008); http://arxiv.org/abs/0908.1884

[4] Elizabeth R. Chen, Michael Engel, and Sharon C. Glotzer, Dense crystalline dimer packings of regular tetrahedra. *Discrete & Computational Geometry* **44**, 253–280 (2010); http://arxiv.org/abs/1001.0586v2

[5] David V. Chudnovsky and Gregory V. Chudnovsky, Approximations and complex multiplication according to Ramanujan. In: George E. Andrews, Richard A. Askey, Bruce C. Berndt, Kollagunta G. Ramanathan, and Robert A. Rankin

(editors), *Ramanujan Revisited*, pp. 375–396; 468–472. Academic Press, Boston (1988)
[6] John H. Conway and Salvatore Torquato, Packing, tiling, and covering with tetrahedra. *Proceedings of the National Academy of Sciences* **103**(28), 10612–10617 (2006); http://www.pnas.org/content/103/28/10612.full.pdf
[7] George Csicsery, *Hard Problems. The Road to the World's Toughest Math Contest* (film). Mathematical Association of America, 2008, 82 minutes (feature) / 45 minutes (classroom version), ISBN 978-088385902-5; http://www.zalafilms.com/hardproblems/hardproblems.html
[8] Simon Gravel, Veit Elser, and Yoav Kallus, Upper bound on the packing density of regular tetrahedra and octahedra. *Discrete & Computational Geometry*, to appear. http://arxiv.org/abs/1008.2830 (online publication: October 14, 2010)
[9] Amir Haji-Akbari, Michael Engel, Aaron S. Keys, Xiaoyu Zheng, Rolfe G. Petschek, Peter Palffy-Muhoray, and Sharon C. Glotzer, Disordered, quasicrystalline and crystalline phases of densely packed tetrahedra. *Nature* **462**, 773–777 (2009)
[10] Thomas C. Hales, A proof of the Kepler conjecture. *Annals of Mathematics* **162**, 1063–1183 (2005)
[11] Jacques Hannak and Emanuel Lasker, *Biographie eines Schachweltmeisters*. Siegfried Engelhardt Verlag, Berlin (1952); third edition (1970). (The quote from the Einstein preface is from http://philosophyofscienceportal.blogspot.com/2008/05/einstein-laskers-and-chess.html)
[12] Martin Henk and Günter M. Ziegler, Spheres in the computer — the Kepler conjecture. In: Martin Aigner and Ehrhard Behrends (editors), *Mathematics Everywhere*, pp. 143–164. American Mathematical Society, Providence (2010)
[13] David Hilbert, Mathematical problems. *Bulletin of the American Mathematical Society* **8**, 437–479 (1902); reprinted: *Bulletin of the American Mathematical Society (New Series)* **37**, 407–436 (2000)
[14] Yoav Kallus, Veit Elser, and Simon Gravel, Dense periodic packings of tetrahedra with small repeating units. *Discrete & Computational Geometry* **44**, 245–252 (2010); http://arxiv.org/abs/0910.5226
[15] Jeffrey C. Lagarias, Bounds for local density of sphere packings and the Kepler conjecture. *Discrete & Computational Geometry* **27**, 165–193 (2002)
[16] Richard Preston, The mountains of Pi. *The New Yorker.* http://www.newyorker.com/archive/1992/03/02/1992_03_02_036_TNY_CARDS_000362534 (March 2, 1992)
[17] Majorie Senechal, Which tetrahedra fill space? *Mathematics Magazine* **54**, 227–243 (1981)
[18] George G. Szpiro, *Kepler's Conjecture. How Some of the Greatest Minds in History Helped Solve One of the Oldest Math Problems in the World*. John Wiley & Sons, Hoboken (2003)
[19] Salvatore Torquato and Yang Jiao, *Analytical constructions of a family of dense tetrahedron packings and the role of symmetry*. Preprint. http://arxiv.org/abs/0912.4210, 16 pages (December 21, 2009)
[20] Eric W. Weisstein, Pi formulas. From MathWorld — A Wolfram Web Resource. http://mathworld.wolfram.com/PiFormulas.html
[21] Oliver Wolf, *The 50th International Mathematical Olympiad 2009* (film). Bildung und Begabung e.V., 48 minutes, Audio Flow 2010; http://www.audio-flow.de
[22] Alexander J. Yee and Shigeru Kondo, 5 trillion digits of Pi — new world record. http://www.numberworld.org/misc_runs/pi-5t/details.html

Complex Dynamics, the Mandelbrot Set, and Newton's Method — or: On Useless and Useful Mathematics

Dierk Schleicher

Abstract. We discuss the theory of iterated polynomials, which is motivated because it is rich, beautiful, and interesting, but not primarily because it is useful. We then discuss the dynamics of the Newton method for finding roots of smooth functions, which is most useful. And finally, we show that they are closely related, and that work on the useful aspects requires deep knowledge of the 'useless' theory. This is an appeal against disintegrating mathematics (or science at large) into 'useful' and 'useless' parts.

Preamble. During the IMO 2009, I had many discussions with contestants. A frequent topic was that students told me that they were especially interested in mathematics and wanted to study it, but they had been told by parents or others that they should rather study a more "useful" field of science, a field with greater chances of getting a good job. This is a concern that I often hear as well from international students on our own campus. In this text, I will try to convey some of my personal answers to these questions.

The first piece of answer I learned as a high school student during the training camp for the German team to the IMO 1983, when I wasn't sure what I would study myself; I found it difficult to choose between physics, computer science, electrical engineering, and mathematics. One of my teachers told me that "smart people will always be needed" no matter what they study. I am convinced that one can be really successful only in a field that one enjoys the most: one can develop one's maximal creativity and potential only in a field that one is excited about. The world is full of opportunity for the creative people, so they can afford choosing their fields of study in terms of maximizing their potential achievements (or their enjoyment!), rather than having to minimize the risk of going unemployed: essentially all the students

Dierk Schleicher
Jacobs University, Postfach 750 561, D-28725 Bremen, Germany.
e-mail: dierk@jacobs-university.de

that I have met in IMO circles, or as math students on our campus, have been quite successful in their future careers, even though they have chosen very different career paths: this means they have had lots of different options for their further careers. (In my own case, I decided to study physics and computer science, only to discover, after earning my Master's degree, that for me the most interesting questions in both fields were mathematical questions; as a result, I turned [back?] into a mathematician and obtained my PhD in this field.)

1 Iteration of Complex Polynomials

Let us start by discussing some (apparently) most useless part of mathematics: iteration of polynomials. Let $q\colon \mathbb{C} \to \mathbb{C}$ be a polynomial of degree at least 2. We are interested in the behavior of q under iteration: i.e., given some $z \in \mathbb{C}$, what will be the long-term behavior of the sequence z, $q(z)$, $q(q(z))$, $q(q(q(z)))$, ...? Let us write $q^{\circ 0} := \mathrm{id}$ and $q^{\circ n} := q \circ q^{\circ (n-1)}$ for the n-th iterate of q. The sequence $(q^{\circ n}(z))_{n \in \mathbb{N}}$ is called the *orbit* of z. Some orbits will certainly be bounded, for instance for those z that are fixed points or periodic points of q (i.e., points z with $q(z) = z$ or $q^{\circ n}(z) = z$ for some $n \in \mathbb{N}$). Other orbits will be unbounded: if $|z|$ is sufficiently large, then $|q(z)| > 2|z|$ and thus $q^{\circ n}(z) \to \infty$ as $n \to \infty$. An important goal in the theory of dynamical systems is to find *invariant sets*, i.e. sets $K \subset \mathbb{C}$ with $q(K) \subset K$; different invariant sets give different answers to the question of what possibilities for the dynamics of the various orbits there are. Two obviously invariant non-empty sets are the *filled-in Julia set*[1] of q

$$K(q) := \{z \in \mathbb{C}\colon \text{the orbit of } z \text{ is bounded under iteration of } q\}$$

and the *escaping set* of q

$$I(q) := \{z \in \mathbb{C}\colon \text{the orbit of } z \text{ tends to } \infty \text{ under iteration of } q\}\ ;$$

we have $\mathbb{C} = K(q) \,\dot\cup\, I(q)$ for every q. Several examples of sets $K(q)$ are displayed in Figure 1.

These sets have a rich and difficult structure; for instance, some sets $K(q)$ are connected, while others are not. Understanding them is not easy and raises a number of questions, some of them quite deep.

---

[1] These sets are named after Gaston Julia (1893–1978), one of the founders of the field of complex dynamics in the early 20-th century; the other major pioneer was Pierre Fatou (1878–1929).

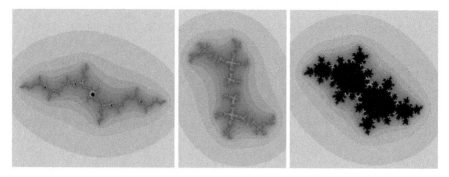

Fig. 1. For several quadratic polynomials q, the filled-in Julia set $K(q)$ is shown in black; the colored complement is the escaping set $I(q)$, where different colors indicate how fast points converge ("escape") to ∞. In some pictures, the black Julia set is so "thin" that it is hardly visible: in the left (and right) pictures, the Julia set is connected, and in the middle picture, it is totally disconnected.

These questions include the following:

(P1) Does $K(q)$ have interior points (i.e., does $K(q)$ contain open sets)? Can the dynamics on open subsets of $K(q)$ be described explicitly?

(P2) How can the different sets $K(q)$ be distinguished topologically or combinatorially?

(P3) Are all disconnected sets $K(q)$ homeomorphic to each other (i.e., topologically the same)?

(P4) If $K(q)$ is connected, is its boundary (the *Julia set*) a curve (i.e., the continuous, not necessarily injective image of a circle)?

(P5) Is it possible that the boundary of $K(q)$ has positive measure? In particular, if $K(q)$ has no interior, can it still have positive measure? (Here we speak of planar Lebesgue measure; the question is thus whether we obtain sets without interior that have positive area.)

(P6) For which (quadratic) polynomials q is $K(q)$ connected (or equivalently, is $I(q)$ simply connected)?

Some questions can be answered relatively easily, others are very recent breakthroughs, and yet others are still open. For instance, (P3) has a simple answer at least in the case of degree 2: if the filled-in Julia set of a quadratic polynomial is disconnected, then it is a *Cantor set*, i.e., it is compact, totally disconnected (every connected component is a point), and it has no isolated points; and any two Cantor sets that are subsets of a metric space are homeomorphic to each other. Moreover, the dynamics on two quadratic Cantor sets is the same (technically speaking, it is "topologically conjugate"). For polynomials of higher degrees, the situation is somewhat more involved, but in light of a very recent theorem all disconnected Julia sets can be described in terms of Cantor sets, as well as connected Julia sets of polynomials of lower degrees. The most interesting Julia sets are thus those that are connected.

Question (P1) also has simple aspects: if $K(q)$ has interior, then any connected component of this interior is called a *Fatou component*, and each Fatou component maps by q onto some other Fatou component. By a deep theorem of Sullivan (conjectured by Fatou before 1920, but proved only around 1980), each Fatou component is either periodic (i.e., it maps onto itself after finitely many iterations), or at least it will map to a periodic Fatou component after finitely many iterations. For a periodic Fatou component $U \subset K(q)$, say of period $n \in \mathbb{N}$, there are only a few possibilities (by a theorem of Fatou):

(A) every orbit in U converges to a periodic point $p \in U$ under iteration of $q^{\circ n}$ (in this case, the orbit of p is called *attracting*);

(P) every orbit in U converges to a periodic point $p \in \partial U$ under iteration of $q^{\circ n}$ (in this case, the periodic orbit of p is called *parabolic*); or

(S) there is a periodic point $p \in U$, and (after a change of coordinates) the dynamics of $q^{\circ n}$ on U is the rotation of a disk by an irrational angle (such components U are called *Siegel disks*, and their existence is a deep theorem originally due to the number theorist Carl Ludwig Siegel; this is discussed in Yoccoz' article [12] in this book).

Question (P5), whether the boundary of $K(q)$ can have positive measure in the plane, has been a deep open question for several decades; in a recent breakthrough, Xavier Buff and Arnaud Chéritat proved that the boundary of $K(q)$ can indeed have positive measure; the "hunt" for the solution is described in [1]. We will get back to this later.

We have seen above that the most interesting Julia sets are those that are connected. How can we decide whether a particular polynomial q has connected filled-in Julia set $K(q)$? It turns out that this is determined by the critical points of q: these are points $z \in \mathbb{C}$ with $q'(z) = 0$. The theorem is the following: *a polynomial q has connected Julia set if and only if the orbits of all critical points of q are bounded (and it is a Cantor set at least if all critical orbits converge to ∞ under iteration)*.

The simplest (non-trivial) case is that of quadratic polynomials. In appropriate coordinates, each of them can be written as $z \mapsto q_c(z) = z^2 + c$ for a unique complex parameter c. For these polynomials, the only critical point (in $\mathbb{C}$) is $z = 0$, so all we need to do is test whether the orbit of 0 is bounded or not: if the orbit is bounded, then $K(q)$ is connected, and otherwise it is a Cantor set. (Background and further reading on the dynamics of polynomials can be found in the excellent book of Milnor [5].)

For $q_c(z) = z^2 + c$, the set of parameters c for which $K_c := K(q_c)$ is connected is called the *Mandelbrot set* $\mathcal{M}$: each point in $\mathcal{M}$ represents a different quadratic polynomial with a different Julia set. In this sense, $\mathcal{M}$ can be viewed as a "table of contents" in the book of all (connected) quadratic Julia sets. With its help, questions such as (P2) or (P4) can be systematically investigated: while there is serious progress, some questions still remain open.

The Mandelbrot set $\mathcal{M}$ itself has a very complicated structure; see Figure 2. Among the open questions are the following:

(M1) Is there a simple way to describe the topology of the Mandelbrot set?
(M2) Is it true that, for every c in the interior of $\mathcal{M}$, the filled-in Julia set of the polynomial $z \mapsto q_c(z) = z^2 + c$ has interior?
(M3) Is the boundary of $\mathcal{M}$ a curve? Does the boundary have measure zero?

These questions are deep and difficult, and not completely answered yet. Question (M2) is often referred to as *"is hyperbolic dynamics dense in the space of quadratic polynomials?"*, and is one of the most important questions in (complex) dynamics. The first half of Question (M3), whether the boundary of $\mathcal{M}$ is a curve, is usually referred to as *"is the Mandelbrot set locally connected?"*, and a positive answer would imply Question (M2) by a fundamental theorem of two of the pioneers, Adrien Douady and John Hubbard [2]. If this was true, then there would be a relatively simple answer to Question (M1): under the assumption that $\mathcal{M}$ is locally connected, there is a simple way to describe the topology of $\mathcal{M}$, the so-called *pinched disk model* developed by William Thurston and Adrien Douady (see for instance [11] and its appendix).

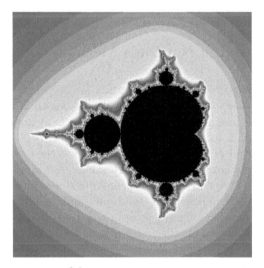

Fig. 2. The Mandelbrot set $\mathcal{M}$ (in black) in the space of iterated complex polynomials $z \mapsto z^2 + c$. Each point represents a different parameter $c \in \mathbb{C}$. Points outside of $\mathcal{M}$ are colored; different color shades describe how fast the point 0 converges to ∞ under iteration of $z \mapsto z^2 + c$.

What we have described are some of the fundamental questions that are investigated in the field of *complex dynamics* (or *holomorphic iteration theory*). These questions are quite theoretical. Why are they interesting, and why useful? After all, a number of smart mathematicians enjoy working on these questions, and they all have their own reasons. Some of my personal answers are:

- these questions arise quite naturally, and they lead to a deep and beautiful theory;
- these questions are related to deep questions in other areas of mathematics, such as number theory and physics (compare for instance Yoccoz' text [12] in this volume), as well as (hyperbolic) geometry, topology, and others;
- while these are phrased for rather special mappings (quadratic polynomials), many of their answers extend to much more general settings (such as iterated polynomials of higher degrees — and more; see below), so that quadratic polynomials serve as prototypes for more complicated situations (compare again Yoccoz' text);
- iteration itself appears in many contexts (even the proofs of some of the most fundamental theorems in mathematics, such as the Implicit Function Theorem, or the Existence Theorem of Ordinary Differential Equations, use an iteration procedure; moreover, many algorithms are of an iterative nature); interesting new insights are often obtained in simple settings.

In my opinion, all of these are valid answers. In Section 3, we will see another reason that gives additional relevance to the iteration theory of polynomials.

2 The Dynamics of Newton's Method

Let us discuss a situation in which iteration is very natural and important: the dynamics of Newton's method. Consider a smooth function $f\colon \mathbb{R} \to \mathbb{R}$, for instance a polynomial. A frequent question in many areas of mathematics and other sciences is to find zeroes of f, i.e., points $x \in \mathbb{R}$ with $f(x) = 0$. Solving any equation $f(x) = g(x)$ can be reduced to the form $f(x) - g(x) = 0$, so this is one of the most fundamental questions of mathematics.

In most cases, there is no explicit formula to find zeroes of f, so the best one can do is to find approximate solutions (and even if there are explicit formulas, it is often more efficient to approximate the solutions). One of the oldest and best known methods is the *Newton method*: given some initial guess x_0, draw a tangent to f at x_0 and see where it intersects the x-axis. This point of intersection, say x_1, is often a better approximation to the true zero than x_0; and this procedure can be iterated to find a sequence of approximations $x_n = N_f(x_{n-1})$. We thus have $x_n = N_f^{\circ n}(x_0)$: based on the point x_0, iteration of the Newton map N_f yields a sequence (x_n) of approximations. In many cases, this sequence converges very rapidly to a zero of f. More precisely, the following theorem is classical and quite basic. *If x^* is any simple root of f (i.e., $f(x^*) = 0$ and $f'(x^*) \neq 0$), then all starting points sufficiently close to x^* have Newton orbits that converge to x^*: there is an $\varepsilon > 0$ so that all x_0 with $|x_0 - x^*| < \varepsilon$ have the property that $x_n \to x^*$.*

Complex Dynamics, the Mandelbrot Set, and Newton's Method

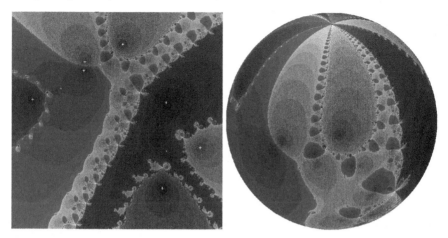

Fig. 3. Left: the dynamical plane of the Newton map for a typical complex polynomial p (here of degree 7). Different colors describe which points in $\mathbb{C}$ converge to which root of p under iteration of N_p. Right: the same Newton dynamics on the Riemann sphere (that is the image sphere of the complex plane union ∞ under stereographic projection); the point at ∞ stands out near the top (where all basins meet).

The same "usually" holds even for non-simple zeroes, but there are some "pathological" counterexamples. In the case of a simple root the convergence is even extremely rapid: in the long run, the number of correct decimal digits in x_n roughly doubles in each iteration step of N_f. For instance, after 10 iterations of N_f, one could expect some $2^{10} > 1000$ valid decimal digits; of course, in practical implementations, the numerical precision would cease to suffice long before that: this means that in fewer than 10 iterations, the approximate answer would be found within the possible numerical precision — provided one has started sufficiently near x_0.

It is easy to give a formula for N_f:

$$N_f(x) = x - f(x)/f'(x).$$

Let us now consider the fundamental case that $f = p$ is a polynomial in one variable. In this case, $N_p(x) = (xp'(x) - p(x))/p'(x)$ is a rational function (the quotient of two polynomials). It may not be so clear why one would want to iterate polynomials, but *Newton maps of polynomials want to be iterated!*

In Figure 3, the dynamics of the Newton map of a typical polynomial is depicted. The formula easily extends from real to complex numbers, and here (as well as in many other contexts), it is much more convenient to work with complex numbers: it often turns out that

Real mathematics is difficult, but complex mathematics is beautiful!

The pictures show that most starting points in $\mathbb{C}$ converge to some root of the polynomial under iteration of N_p. Some natural (and important!) questions include the following:

(N1) Is it true that *almost all* points in $\mathbb{C}$ converge to some root of p under iteration of N_p? (Will a random starting point converge to some root with full probability? Or equivalently, does the set of starting points that do not converge to any root have measure zero in the plane?) This would be the *best possible* case.

(N2) Is it possible that there are *open sets* of points in $\mathbb{C}$ none of which converge to any root of p under iteration of N_p? This would be *worst possible* case.

(N3) Suppose all that one knows is that some polynomial p of degree $d \geq 2$ has all its roots in the complex unit disk $\mathbb{D}$. How can one find points at which to start the Newton iteration in order to be sure that one finds *all* roots of p?

(N4) Starting at appropriate points, how many times does one have to iterate in order to find all the roots with a given precision $\varepsilon > 0$?

For any root α of p, let B_α be the *basin* of α: this is the set of all $z \in \mathbb{C}$ that converge to α under iteration of N_p. It is easy to see that each basin is open, and it contains a neighborhood of α. The union of the basins are the "good" starting points (starting the Newton iteration there will find some root); their complement are the "bad" starting points (starting the iteration there will not lead to any root).

For a quadratic polynomial p with two distinct roots, it is quite easy to see that the set of bad starting points always forms a single straight line symmetric to the two roots. The interesting cases happen when the polynomial p has degree $d \geq 3$. First of all, observe that every Newton map (for a polynomial p of degree $d \geq 2$) must have bad starting points z: for instance, there are always periodic points of periods $n \geq 2$ (but their number is always countable). Moreover, the boundary of any basin is a closed non-empty set, and it cannot intersect the basin of any root. Therefore, all points in all basin boundaries cannot converge to any root. (Interestingly, it turns out that the boundaries of *all* basins always coincide! This common boundary is the *Julia set* of N_p.) Questions (N1) and (N2) can thus be rephrased as follows: if a point $z \in \mathbb{C}$ does not converge to any root of p, does this imply that z is in the common boundary of all basins? And can this common boundary have positive measure?

3 The Useless and the Useful

Clues to some of these questions can be found in Figure 4: for certain cubic polynomials, the set of bad starting points contains subsets that resemble filled-in Julia sets of quadratic polynomials. This is explained by a fundamental theory of Adrien Douady and John Hubbard of "polynomial-like maps" and "renormalization": *for every quadratic polynomial q, there is a cubic polynomial p so that in the Newton dynamics N_p, there is a copy of the filled-in Julia set of q within the set of bad starting points.* In fact, in a precise sense, *most* bad starting points belong to such small copies of the filled-in Julia sets of quadratic polynomials (all the others have measure zero). Therefore, an understanding of the bad starting points for the Newton dynamics needs an understanding of the dynamics of iterated polynomials! *The "useful" questions of Newton dynamics requires knowledge of the "useless" theory of iterated polynomials!*

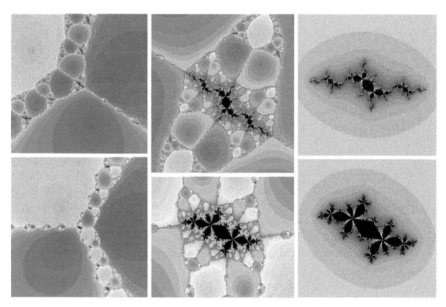

Fig. 4. Newton maps for two cubic polynomials (left), with magnifications of some details (center). Black are "renormalizable" starting points that do not converge to any root at all; for starting points that do converge to a root, colors describe to which root they converge. Right: Filled-in Julia sets for two quadratic polynomials: these are homeomorphic to (a connected component of) the set of black points in the center pictures.

It follows from the renormalization theory that the answer to question (N1) is *no:* there are polynomials p so that the boundary of set of points converging to some root has *positive measure* in the plane, because this boundary contains a copy of the boundary of the filled-in Julia set of a quadratic polynomial with positive measure: this uses the recent result on question (P5). In such cases, the set of bad starting points may or may not have interior.

Worse yet, since there are many quadratic polynomials q whose filled-in Julia sets do have interior, there are many cubic polynomials p so that the bad starting points for the Newton map N_p have interior, i.e., they contain open sets; this gives an answer to question (N2). A few examples are shown in Figure 4.

The "worst-case scenario" that Newton maps may have open sets of bad starting points was discovered, rather unexpectedly, in the late 1970's, after systematic computer experiments became available (pioneered by John Hubbard). This led to the following question that was raised by Fields medallist Stephen Smale [8, Problem 6] and others:

(N5) Classify all polynomials p (of all degrees) so that the corresponding Newton maps have open sets of bad starting points.

Some 25 years later, we are now ready to answer this question. Some form of the answer is contained in the forthcoming PhD thesis of Yauhen Mikulich in Bremen (himself a winner of a First Prize at the International Mathematics Competition for University Students). We skip the details, but roughly speaking he shows that an understanding of the useful and important question (N5) involves a detailed understanding of the "useless" dynamics of iterated polynomials of all degrees! In all degrees $d \geq 3$, all the "bad" cases are contained in copies of little filled-in Julia sets of polynomials q of some degrees (up to sets of measure zero), and their classification requires a classification of all iterated polynomials.

Let us investigate a specific "toy-model" case of low degrees. We saw above that Newton maps of quadratic polynomials are simple. But Newton maps of cubic polynomials are already much more interesting. So let us consider a cubic polynomial $p(z) = c(z-\alpha_1)(z-\alpha_2)(z-\alpha_3)$ with $c, \alpha_1, \alpha_2, \alpha_3 \in \mathbb{C}$. The factor c is irrelevant for the Newton map $N_p(z) = z - p(z)/p'(z)$, so we can set $c = 1$. Translating coordinates, we may assume $\alpha_1 = 0$, and rescaling and possibly relabeling, we may assume that $\alpha_2 = 1$ (unless $\alpha_3 = \alpha_2 = \alpha_1 = 0$). In convenient coordinates, every cubic polynomial other than z^3 may thus be written as $p_\lambda(z) = z(z-1)(z-\lambda)$, and the Newton dynamics respects these coordinates. Every $\lambda \in \mathbb{C}$ represents some cubic polynomial with its own Newton map $N_{p_\lambda} =: N_\lambda$.

The investigation starts with the following classical theorem: *if, for the Newton map of a cubic polynomial p, the set of "bad" starting points contains an open set, then it also contains the center of gravity of the three roots of p.*

In order to test whether the Newton map of a given cubic polynomial has open sets of bad starting points, it suffices to iterate a single point (which happens to be be the unique point $z \in \mathbb{C} \setminus \{\alpha_1, \alpha_2, \alpha_3\}$ with $N'_\lambda(z) = 0$: it is the only "free critical point" of N_λ; this is analogous to the behavior of critical points that we saw earlier). In Figure 5, the λ-plane of the cubic polynomials p_λ is depicted, and all parameters λ are colored black for which the free critical point does not converge to any root. It turns out that an understanding of the "bad polynomials" even in the simple cubic case involves an understanding of the Mandelbrot set! (See [9].) Of course, an understanding of the general case of higher-degree polynomials

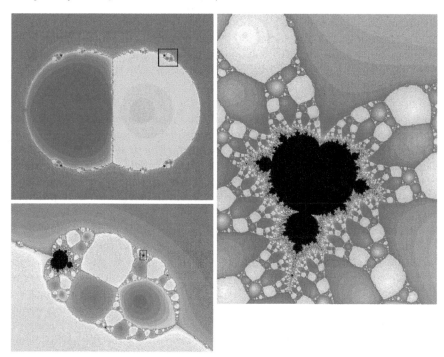

Fig. 5. The λ-plane of cubic polynomials $p_\lambda(z) = z(z-1)(z-\lambda)$ for $\lambda \in \mathbb{C}$. Black points denote those polynomials for which the free critical point does not converge to any root. Shown are a large-scale view of the λ-plane (upper left) and two subsequent magnifications (lower left and right). Colors distinguish which root the free critical point converges to.

involves an understanding of higher-dimensional analogues of the Mandelbrot set, and this is much more complicated, but the basic principle is the same.

4 Old Questions and New Answers

Let us come to some very practical questions. Newton's method was designed to find zeroes of smooth functions. In the fundamental case of complex polynomials of a single variable, how does one actually find all the roots? Newton's method, as described above, is a heuristic principle: choose a starting point somewhere, start iterating the Newton map and hope the orbit converges to some root. Even if almost all starting points converge to some root: how can one make sure that all roots are found? Is it conceivable that some root is "hiding" somewhere and can be found starting only at a small set of starting points? (Of course, when some roots are found, one could in principle factor them out and start over with a polynomial of lower degree. But in practice, this is often not an option: deflation of polynomials is in general numerically very unstable; moreover, if a polynomial has a special and easy-to-evaluate form, then this may no longer be so after deflation. We thus want to find *all* roots without deflation.)

One of our goals is to turn Newton's method into an algorithm. We want a recipe of the following kind: *given a polynomial p of some degree $d \geq 2$ and a desired accuracy $\varepsilon > 0$, take the following starting points $z^{(1)}, \ldots, z^{(k)}$ with $k \geq d$ (specify) and iterate the Newton map starting at these points until the following condition is satisfied (specify); then the following d points are ε-close to the d roots of p (specify).* (Of course, the starting points $z^{(j)}$ must be specified explicitly; in our approach, they do not depend on p at all, provided p is appropriately normalized.)

Newton's method is as old as analysis, but there is no complete theory known about it, and it is not (yet) an algorithm. One of the main problems is that an orbit $z_n = N_p^{\circ n}(z_0)$ could reach a point z where $p'(z)$ is very close to zero, so that $N_p(z) = z - p(z)/p'(z)$ is near ∞, and from then on it would take a long time until the orbit gets back to where the roots are: one thus loses control on the dynamics. It is difficult to predict for which starting points this happens. Therefore, numerical analysis contains an elaborate theory on polynomial root-finding, but Newton's method has a reputation as being difficult to control, so most of the theory is on different numerical methods. A recent overview on what is known on Newton's method as a root-finder can be found in [6].

However, methods from complex dynamics, and from the "useless" parts of the theory allow us to make progress on the practical task of turning Newton's method into an algorithm. For instance we have the following theorem [4]: *given a (suitably normalized) polynomial of some degree $d \geq 2$, one can specify a relatively small set of $k = \lceil 1.11 \, d \log^2 d \rceil$ starting points $z^{(1)}, z^{(2)}, \ldots, z^{(k)}$, so that for each root α of p there is at least one of these starting points that converge to α under iteration of N_p.* This is a set of "good starting points", and it depends only on the degree d, not on the specific polynomial p (provided it is suitably normalized). This is a good answer to question (N3) raised above. (This set of starting points is easy to write down; these points are equidistributed on $\log d$ concentric circles around the origin.)

This is a good beginning for turning Newton's method into an algorithm, but it is not quite complete. One of the most important remaining questions is (N4), how many iterations does it require until all the roots are found with prescribed precision ε? Here, too, there is progress in sight, using methods from complex dynamics, in particular using an interplay between Euclidean and hyperbolic geometry; it is indeed possible to give explicit upper bounds on how many iterations are required in order to find all the roots, and these bounds are not too bad (this is current work in progress [7]). Newton's method may be a much better algorithm than previously thought — this is one of the oldest questions of analysis, and yet there is still a lot of work that one

Fig. 6. The dynamics for the Riemann ξ function: some of the zeroes are marked in the picture, and different colors distinguish to which zero of ξ any point in the plane converges under Newton's method.

can do. Contrary to certain preconceptions, even old parts of mathematics still leave room for new discoveries — often exactly because there is progress in other areas, whether or not they are deemed useful ahead of time. (This point of view is also mentioned in the article by Timothy Gowers [3].)

We cannot resist the temptation to mention the case of functions f that are more general than polynomials, for instance "entire functions" $f\colon \mathbb{C} \to \mathbb{C}$ that are (complex) differentiable everywhere in $\mathbb{C}$. A most prominent example is the famous "Riemann zeta function" ζ as discussed in the article by Terence Tao [10]: it has zeroes at the negative even integers, and the Riemann conjecture says that all remaining zeroes have real parts $1/2$. The ζ function is not an entire function (it has a pole at $z = 1$), but it has a close cousin, the ξ-function: its zeroes are exactly the "non-trivial" zeroes of ζ. To this function, one can apply Newton's method as well: see Figure 6. The location of these zeroes is of fundamental importance for many results in mathematics, and Newton's method is designed for finding these roots. Of course, there are special methods designed specifically for the ζ function, but Newton's method is general enough to deal even with such functions; we have to omit the details.

Let me conclude with a personal note. I believe there is no useful or useless mathematics. Different areas of mathematics can be more or less interesting — that is a matter of personal taste. But there are no areas of mathematics that should be classified as "useful" or "useless". Mathematics is full with connections and interrelations, some of them obvious, while others are discovered only after a long time. Some of us are motivated by the intrinsic beauty of our mathematical areas, others because we can solve questions that were raised by others, inside mathematics or outside. We are all building the same house, and what matters is the willingness to observe new interrelations when they arise, the willingness to think more broadly beyond narrow borders of sub-disciplines, and the willingness to go more deeply into uncharted territory. Good mathematics may find its applications by itself, sooner or later, and quite possibly in unexpected ways. Restricting the paths of discovery only to obvious applications would miss some of the most important relations — it would be a waste of those talents that are at their best when building up a good theory!

References

[1] Arnaud Chéritat, The hunt for Julia sets with positive measure. In: Dierk Schleicher (editor), *Complex Dynamics: Families and Friends*, pp. 539–559. AK Peters, Wellesley (2009)

[2] Adrien Douady and John Hubbard, Etude dynamique des pôlynomes complexes (the "Orsay Notes"). *Publications Mathématiques d'Orsay* 84-02 (1984) and 85-04 (1985)

[3] W. Timothy Gowers, How do IMO problems compare with research problems? Ramsey theory as a case study. In: Dierk Schleicher and Malte Lackmann (editors), *An Invitation to Mathematics: From Competitions to Research*, pp. 55–69. Springer, Heidelberg (2011)

[4] John H. Hubbard, Dierk Schleicher, and Scott Sutherland, How to find all roots of complex polynomials by Newton's method. *Inventiones Mathematicae* **146**, 1–33 (2001)

[5] John Milnor, *Dynamics in One Complex Variable, third edition.* Princeton University Press, Princeton (2006)

[6] Johannes Rückert, Rational and transcendental Newton maps. In: Mikhail Lyubich and Michael Yampolsky (editors), *Holomorphic Dynamics and Renormalization. A Volume in Honour of John Milnor's 75th Birthday*, Fields Institute Communications, volume 53, pp. 197–212. American Mathematical Society, Providence (2008)

[7] Dierk Schleicher, Newton's method as a dynamical system: efficient root finding of polynomials and the Riemann ζ function. In: Mikhail Lyubich and Michael Yampolsky (editors), *Holomorphic Dynamics and Renormalization. A Volume in Honour of John Milnor's 75th Birthday*, Fields Institute Communications, volume 53, pp. 213–224. American Mathematical Society, Providence (2008)

[8] Stephen Smale, On the efficiency of algorithms of analysis. *Bulletin of the American Mathematical Society (New Series)* **13**(2), 87–121 (1985)

[9] Tan Lei, Branched coverings and cubic Newton maps. *Fundamenta Mathematicae* **154**, 207–260 (1997)

[10] Terence Tao, Structure and randomness in the prime numbers. In: Dierk Schleicher and Malte Lackmann (editors), *An Invitation to Mathematics: From Competitions to Research*, pp. 1–7. Springer, Heidelberg (2011)

[11] William Thurston, On the geometry and dynamics of iterated rational maps. Manuscript (1982). In: Dierk Schleicher (editor), *Complex Dynamics: Families and Friends*, pp. 3–137. AK Peters, Wellesley (2009)

[12] Jean-Christophe Yoccoz, Small divisors: number theory in dynamical systems. In: Dierk Schleicher and Malte Lackmann (editors), *An Invitation to Mathematics: From Competitions to Research*, pp. 43–54. Springer, Heidelberg (2011)